THE
BIG QUESTIONS
SCIENCE

THE
BIG QUESTIONS
SCIENCE
OF

ANTONINO DEL POPOLO

The University of Catania, Italy

World Scientific

NEW JERSEY · LONDON · SINGAPORE · BEIJING · SHANGHAI · HONG KONG · TAIPEI · CHENNAI

Published by

World Scientific Publishing Co. Pte. Ltd.

5 Toh Tuck Link, Singapore 596224

USA office: 27 Warren Street, Suite 401-402, Hackensack, NJ 07601

UK office: 57 Shelton Street, Covent Garden, London WC2H 9HE

British Library Cataloguing-in-Publication Data
A catalogue record for this book is available from the British Library.

THE BIG QUESTIONS OF SCIENCE

ISBN 978-981-12-9407-5 (hardcover)
ISBN 978-981-12-9408-2 (ebook for institutions)
ISBN 978-981-12-9409-9 (ebook for individuals)

For any available supplementary material, please visit
https://www.worldscientific.com/worldscibooks/10.1142/13864#t=suppl

Desk Editor: Rhaimie Wahap

Typeset by Stallion Press
Email: enquiries@stallionpress.com

About the Author

Antonino Del Popolo is a researcher and professor of astrophysics and cosmology at the Department of Physics and Astronomy at Catania University in Italy. His main interests are physical cosmology, dark matter, dark energy, extrasolar planets, and exobiology. He is the author of more than 200 papers published in top scientific journals, including *Nature*. He has served as a visiting professor and given numerous talks at many institutes and universities worldwide.

Contents

Chapter 1

Introduction

The history of astronomy is a history of receding horizons.

— Edwin Hubble

In *Antigone*, Sophocles writes about how there are many extraordinary things, but that the most extraordinary of all is man. Where this extraordinary nature comes from is indicated by the etymology of the word *antrophos*, the term used by the Greeks to refer to the human being, which, in the *Cratylus*, is traced back by Plato to the verb *anathrein*, which means "to feel oneself being", that is, one who realises what he sees. This characteristic of being aware of himself and what surrounds him leads him to constantly ask himself questions. As Aristotle wrote in the *Metaphysics*:

> *A sense of wonder leads men to philosophise, in ancient times and even today. Their wonder is initially aroused by trivial things; but they subsequently continue to wonder about less mundane questions such as the changes of the Moon, Sun and stars, and the beginning of the Universe. What is the result of this amazement? A profound feeling of ignorance! Man begins to philosophise, therefore, to emerge from ignorance. All men, by their nature, desire to know.*

This characteristic has remained a typical human trait over the millennia. Time has not changed this peculiarity of man to question what surrounds him. Children are the most evident manifestation of this innate curiosity of human beings. Among the first questions children ask us are

1

where does the world around us come from, who made it, and where do we come from. They are never tired of asking questions, often putting adults in difficulty, adults who are unable to answer many of those questions. There are many questions to which we still do not have an answer: Who are we? Where do we come from? Why does the world exist? What meaning do we and the Universe have? The mystery of existence induces in man a sense of wonder and amazement that leads him to continually question himself about it.

All these questions have led to the birth of myths, they have given rise to cosmogonies, and they have led to the birth of philosophy, religions, and finally science – different ways to try to unravel those questions that have always followed us, different ways of trying to understand reality. Although the existential questions that accompany us in our growth and evolution have persisted, awaiting precise answers, in recent centuries science has given answers to many other questions, made us increasingly aware of the world around us, and allowed us, with an understanding of the laws of nature, to use those laws to our advantage. So, our tendency to ask ourselves existential or non-existential questions has allowed us to evolve and create a technological world that has simplified our lives compared to those of our ancestors. Though we have not been able to give a final answer to many questions, today, we can look around, or look up at the sky, and proudly tell ourselves that part of the Universe is comprehensible to us and that the future will broaden our sphere of understanding and knowledge. If for the Australian Aborigines, the Sun was a woman who woke up in her camp in the east, then lit a torch that she carried around the sky, or for the Indo-European peoples, the Sun and the Moon moved in the sky on chariots pulled by horses and driven by a charioteer, today we know that the Sun is a star, like the others we see in the firmament, a "furnace" supported by nuclear fusion reactions, and the Moon is simply a satellite that revolves around the Earth. If the Vikings believed that solar eclipses originated from a wolf that captured the Sun and mauled it, today we know that eclipses are natural phenomena produced by the Moon's shadow falling on the Earth. We could continue to give many other similar examples. Although today we are more aware than our ancestors, at the same time, we realise that as our knowledge grows, the edge of our ignorance grows simultaneously.

With our knowledge of astronomy, we can today measure the magnitude of our ignorance. We know that the objects observable in our Universe make up only 5% of its total mass, while the remaining 95% is catalogued under the terms *dark matter* and *dark energy*, about which we know little or nothing. However, we are confident about the future, and to paraphrase Edwin Hubble, we know that the history of man *is a history of receding horizons.*

In this book, we show how the simplest questions have led to humanity's greatest discoveries. We seek to fuel readers' curiosity with the deepest questions about our Universe, and our place in it, by providing the answers we have. At the same time, by doing this, we lead you into a maelstrom of uncertainties and gaps in human knowledge.

Chapter 2

Why There is Something Rather Than Nothing?

The universe is the way it is, whether we like it or not.

— Lawrance M. Krauss

The Universe is made up of an enormous quantity of stars, galaxies, and matter visible with our telescopes. As we will see in Chapters 7 and 8, all visible matter is only a small percentage of the mass that makes up the Universe. Most of the material content of the Universe is made up of invisible components that we call dark matter and dark energy. One of the questions that often comes to mind is: Why is there something instead of nothing? Why does the Universe exist? These ancient questions do not yet have an answer, but compared to our ancestors, our current knowledge of the laws of physics allows us to describe the events that, starting from Planck time (10^{-43} s after the origin of the Universe), led the Universe to evolve to its current state, which is full of small structures such as planets, larger ones such as stars and galaxies, and much larger ones such as clusters and superclusters of galaxies. All these originated from what is called a *gravitational singularity*, a point at which physical quantities lose meaning, about 13.8 billion years ago. To intuitively realise that things really happened this way, ideally it is enough to reverse the expansion motion by which the galaxies are animated, as is done when rewinding a film. What we would expect would be to see the galaxies reverse their motion and all start moving towards one point. Going back in time, the Universe would become smaller and hotter, reaching temperatures so high that matter as we know it today did not exist. We would reach a point when the Universe had dimensions of 10^{-35} m, the Planck era, in which the known laws of physics would no longer apply. At smaller scales, a still non-existent theory known among physicists as *quantum gravity* is needed, which is a combination of the theory of gravity and quantum mechanics. Starting with Planck's time, known physics allows us to describe the most important phases of the evolution of the Universe. The matter that makes up the objects we know and the entire Universe origi-nated thanks to that initial event. When the Universe was very young, about three minutes old, the lightest elements such as hydrogen and helium were formed, and going forward in time, after a few hundred mil-lion years from the beginning, the stars that formed scattered the heaviest elements throughout space. What surrounds us has very ancient origins. The atoms that make up our body come from stars that exploded in the ancient past. We are like "cosmic Legos", where each brick comes from different regions of the cosmos. To answer the question, "Why *does some-thing exist instead of nothing*", it is necessary to understand what science tells us about our distant past and how the wonder that surrounds us could be formed from a singularity.

Brief History of the Big Bang Theory

The conception of the cosmos from ancient times to the Middle Ages and beyond was based on Aristotelian physics and Ptolemy's astronomical models. This Universe was static, and nothing changed; the planets and stars moved following immutable and eternal cycles. The idea of stillness died so hard that we had to wait until the 20th century to see real change. The idea of a static Universe was shared by 20th-century scientists, including Einstein (before Hubble's discovery of the expansion of the Universe). Two years after the publication of his treatise on the theory of general relativity, in 1917, Einstein applied it to the Universe and found results that contradicted his prejudices and those of his time: that is, the static nature of the Universe. He discovered that his equations predicted a non-static Universe. Following the generalised prejudice that the Universe was static, he introduced a constant into his equations, the cosmological constant Λ, which, by acting repulsively, counteracted gravitational attrac-tion and meant that the Universe could remain static.

In 1922, the Russian meteorologist and mathematical physicist Friedmann published an article in which he showed that the solutions to

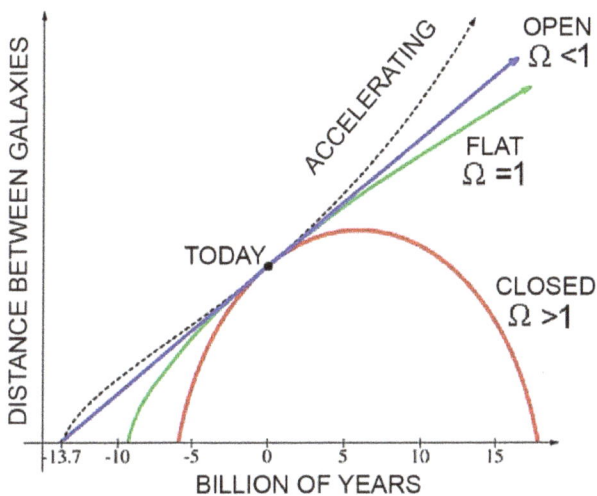

Figure 2.1. Possible behaviours of the universe depending on its density.

the equations of general relativity applied to the Universe provided three different solutions.

The three types of universes, i.e. the solutions to Einstein's equations, depend on the mass-energy density of the universe. The three possible types of universes are shown in Figure 2.1. Each curve represents how far galaxies separate over time. The vertical axis of Figure 2.1 shows the so-called expansion parameter, which represents the relative expansion of the universe and can be considered the average distance between galaxies. On the horizontal axis, we have time. There is a particular value of the density of the universe, called *critical density*, ρ_c, whose value corresponds to approximately 5 hydrogen atoms per cubic metre, which distinguishes the three different types of geometry of the universe. If the average density of the universe, ρ, is greater than the critical density, ρ_c, i.e. if there are more than 5 hydrogen atoms per cubic metre in the universe, the universe expands to a maximum size and then collapses again, as seen in the orange curve of Figure 2.1. The universe begins with a big bang and ends with a *big crunch*, which is the opposite phenomenon of the big bang. In this type of universe, called a closed universe, the structure of space is similar to that of a sphere, and the universe is finite. If the density is equal to the critical density, as shown in the green line in Figure 2.1, the universe will expand forever. The spatial structure will be similar to that of a plane, which is why we talk about a "flat universe". Finally, if the density is less than the critical one, as seen in the blue curve of Figure 2.1, the universe will expand forever, and for this reason we speak of an "open universe", and the spatial geometry resembles that of a saddle horse, and space is infinite. In cosmology, for practical reasons, instead of using the critical density ρ_c, the ratio between the density ρ and the critical density ρ_c is used, which is called the *density parameter* and is indicated by $\Omega = \rho/\rho_c$. The density parameter also expresses the amount of mass of a certain species in terms of the critical density. The three types of geometry that we just discussed and shown in Figure 2.1 can be expressed in terms of the density parameter Ω as follows: the flat universe, with density equal to the critical one, has Ω equal to one ($\Omega = 1$); the open one has Ω less than one ($\Omega < 1$); and the closed one has Ω greater than one ($\Omega > 1$). These are Friedman's classic solutions. Today, we know that the Universe is dominated by *dark energy* (which we will talk about in Chapter 8), and this causes the Universe to expand in an accelerated manner, as shown by the dashed curve in Figure 2.1.

In addition to showing that Einstein's universe was unstable, either expanding or collapsing, Friedman was the first to assert that the Universe would somehow have a birth certificate and also estimated its age at 10–20 billion years. The Universe would have been born from what mathematicians and physicists call a gravitational singularity, that is, a point where the physical quantities, such as density, tend towards an infinite value. Examples of gravitational singularities, in addition to those linked to the birth of the Universe, are the *black holes* (which we will talk about in Chapter 5), which swallow everything that passes within a certain radius, even light!

A gravitational singularity is a point in space where the curvature of space-time is infinite. In other words, it is a point where space and time merge and where physical quantities, such as distance, curvature, and volume, cannot be calculated. In 1970, Stephen Hawking and Roger Penrose were the first to predict the existence of this space-time singularity, starting with the equations of general relativity and showing that, in this theory, the singularity was inevitable. This implies that general relativity cannot be used to study what happened in the period when this singularity was present, which, as we will see, is the so-called Planck era (the period between 0 and 10^{-43} s). A theory that also includes quantum effects is needed. This theory, known as *quantum gravity*, has not yet been fully developed.

For Friedman, however, this singularity was only something mathematical, not physical. Only several years later, in 1927, the Jesuit priest George Lemaitre gave a physical meaning to this gravitational singularity, or, more precisely, to the moments following the singularity: the explosion of the so-called *primaeval atom*, today known as the *big bang*. This term was coined by Fred Hoyle ironically since he was a proponent of a model of the Universe in competition with the Big Bang model: the so-called *steady state theory*.[1] During a BBC broadcast trying to explain the theory, he said, "Those theories have been based on the hypothesis that all the matter in the universe was created in a "big bang" at a particular time in the distant past".

Lemaitre's contributions to astronomy are many. In 1925, he unknowingly found Friedman's solutions to the equations of general relativity. In 1927, Lemaitre published his famous article in an obscure Belgian magazine: "A Homogeneous Universe of Constant Mass and Increasing Radius That Justifies the Radial Velocity of Extragalactic Nebulae".

In this work, he considered the dynamic solutions to the equations of general relativity from a more physical perspective than Friedman's and assumed that the radius of the Universe could vary arbitrarily with time. Among the results of the work was the relationship between the speed of recession, i.e. the away movement, of extragalactic nebulae and the distance to where they are found, now known as the *Hubble–Lemaitre law*.

Although Lemaitre had theoretically determined the law of the expansion of the Universe, this had not been accepted by scientific circles. He himself had contributed to this since Lemaitre himself had deleted almost every reference to it in the English translation of his 1927 article.

Only two years later, in 1929, Hubble published the article "A Relationship Between Distance and Radial Velocity Between Extragalactic Nebulae", in which he showed that there was a direct proportionality relationship between the recession speeds of galaxies and their distances.

In his 1931 article, "The Expanding Universe", Lemaitre argued that a static proto-Universe could exist in which *all the energy would have*

---

[1]The *steady state theory* is a cosmological theory proposed by Fred Hoyle, Herman Bondi, and Thomas Gold, based on the *perfect cosmological principle*, i.e. the idea that the Universe is homogeneous and isotropic in space and time. The Universe is expanding and always has the same properties at any time and in any position. The decrease in density due to expansion is obviated by the continuous creation of matter.

been in the form of electromagnetic radiation which would have immediately condensed into matter.

This idea was confirmed in the article "The Beginning of the World from the Point of View of Quantum Theory" published in *Nature* in 1931, in which he reported the following:

> *We could conceive the principle of the universe in the form of a single atom whose atomic weight is the total mass of the Universe. This highly unstable atom would have split into smaller and smaller atoms during a kind of super-radioactive process.*

In the article "The Expansion of Space", Lemaitre wrote:

> *The expansion thus took place in three phases: a first period of rapid expansion in which the atom-universe broke into atomic stars; a period of slowdown followed by a third period of accelerated expansion.*

Lemaitre's model showed that the Universe had an origin, but his study did not go back to the time of the singularity. He did not discuss the origin of the primordial atom but assumed its existence before the so-called *radioactive explosion*. That is, according to him, it made no sense to discuss the properties of the primordial atom before the instant of the explosion since space and time would only begin after its disintegration. This is exactly the modern point of view: space and time originated with the big bang; therefore, it makes no sense to discuss what came before. Quoting Saint Augustine, *the world was created with time, not in time.*

Hubble's law is the first fundamental evidence underlying the Big Bang theory. This law shows a direct proportionality relationship between the speed at which galaxies move away and the distance. More distant galaxies have faster recession speeds than closer ones. From its discovery to today, this law has been confirmed many times and is not only valid for gravitationally bound groups. For example, the Andromeda Galaxy is moving towards us instead of receding.

There are two other pillars on which this theory is based. The second is the *Cosmic Microwave Background (CMB) Radiation,* discovered in 1965, which is a sort of sea of residual microwaves that pervades the Universe and reaches the Earth from all over the sky. A true imprint of the big bang. It was predicted in 1948 by Alpher, Herman, and Gamow, and its temperature, now known to be 2.725 K, is very close to the value obtained by Alpher and Herman in 1950. The third is the prediction of the abundance of light elements generated in *primordial nucleosynthesis*, obtained with Ralph Alpher's calculations and published in the famous article "αβγ".

Although the theory has the name of the big bang, it was not an explosion like the ones we know, where the explosion starts from a point and propagates outwards. The big bang was not an explosion within the Universe but an expansion of the Universe. Furthermore, the big bang did not happen in one place but happened everywhere. All the dots that make up the Universe today – a pimple on your nose, a dot on Andromeda, and so on – were close enough to touch each other.

A simple example that clarifies the expansion of the Universe and Hubble's law is that of an inflating balloon with dots drawn on it. In the similitude, the dots represent galaxies, and the balloon represents space. When the balloon inflates, the rubber is stretched, and the dots on it move away from each other. If we suppose we are one of the dots, we see all the other dots moving away from us, and we think we are at the centre of the Universe. In reality, we are not standing still but moving away from the other dots. From any dot, the observed situation is identical: the other dots are seen moving away with a speed proportional to the distance. In other words, there is no privileged point; the expansion is identical for every point in the Universe.

Today, most physicists accept the idea that the Universe was born from the big bang, that is, from a *space-time singularity*, a sort of non-place and non-time from which everything around us originated. However, alternative models without singularities exist. For example, cyclic universes characterised by an infinite cycle of creation and destruction have been proposed, such as the *ecpyrotic model*, which we discuss in Chapter 9. James Hartle and Stephen Hawking proposed their model of the origin of the Universe, the so-called *Hartle–Hawking state*, which describes a universe born from nothing, without an initial singularity, an initial state without boundaries. Such a universe would be self-sufficient and self-created. To describe it, the authors used the concept of imaginary time. Although the Big Bang theory is not the only theory that describes the origin of the Universe, it is certainly the one with the greatest amount of supporting data, as well as the most accepted and popular. To answer our question, "Why does something exist instead of nothing?" We must first answer another question, namely, "What is this 'something'?" Or, in other words, "What is the Universe composed of?" We must therefore discuss the characteristics of the primordial Universe and its evolution up to today.

The Origin of Things

About 14 billion years ago, the Universe had infinitesimal dimensions and contained all the matter and energy that constitute it today. As already mentioned, there is a wall beyond which our theoretical knowledge of the Universe stops. This is the Planck wall, which takes its name from the German physicist Max Planck, corresponding to approximately 10^{-43} s^2, characterised by pressures, temperatures (10^{33} K^3), and densities (10^{96} g cm^3) so high that space was distorted, folded, and made up of a foam of mini black holes and wormholes (sort of gravitational tunnels) (see Chapter 6) with dimensions of the order of 10^{-33} cm, temperatures of 10^{32} K, and evaporation times of the order of Planck time. The cosmological horizon of the Universe was approximately 10^{-33} cm. These are conditions so extreme that they cannot be recreated even in the most powerful particle accelerator in the world, the LHC at CERN in Geneva. During Planck time, gravity, which in our world is enormously less intense than other forces, had intensities comparable to those of the other forces, namely the *weak and strong nuclear forces* and *the electromagnetic force*. To describe this era, we need a theory that brings together quantum mechanics and general relativity, the so-called quantum gravity, which has not yet been formulated. Physicists have been trying for decades to find a theory that describes this situation in which all forces were unified, the *Theory of Everything*, usually abbreviated to TOE. There have been attempts to understand what the Universe was like before Planck time (10^{-43} s) without great success. In any case, the results obtained are of dubious validity given that the physics that we know and use loses meaning before Planck's time. In the Planck era, the Universe was dominated by the *quantum vacuum* (see Chapter 5). This vacuum is not what is usually meant by a vacuum, a region devoid of everything, but it is an entity full of energy and liveliness from which particle–antiparticle (or matter–antimatter) pairs are created, which, after infinitesimal times, annihilate and return to the void. The *era of great unification* unfolded between 10^{-43} and 10^{-36} s, at the beginning of which the gravitational force

---

[2]For simplicity, we use scientific notation. Therefore, for example, 10^{-2} indicates the number 0.01, while 10^2 indicates the number 100.

[3]Temperatures are measured in degrees Kelvin, K. K is a unit of temperature measurement called Kelvin, named after William Thomson, also known as Lord Kelvin. To obtain temperatures in degrees Centigrade, °C, simply subtract 273.15 from the Kelvin value.

separated from the other forces, the electroweak and the strong nuclear forces. The strong force and the electroweak force were unified, and they behaved like a single force. The temperature, which at Planck's time was above 10^{30} K, decreased to 10^{28} K towards the end of the era of great unification, and the Universe was subject to the decoupling of the strong interaction from the electroweak force, which constitutes two separate forces. This phase is called the *grand unification phase transition*. Some types of phase transitions are present in everyday life. For example, water cooled to 0°C turns into ice. We all know this phase transition. Furthermore, we move from a situation with higher symmetry to one with lower symmetry. For example, water appears the same from any direction we look at it and has greater symmetry, while ice has a crystalline structure and is not symmetrical like water. Similar to what happens in the water–ice phase transition, where the latent heat is released, which causes the phase transition, an enormous amount of energy was released in the Universe, the energy present in the vacuum. The Universe, of sub-nuclear dimensions, underwent a metamorphosis, thanks to a period of exponential expansion, or *inflation*, which developed in the period between 10^{-36} and 10^{-32} s (see Appendix 2). From dimensions of the order of 10^{-28} metres, the Universe grew by a factor of 10^{25}–10^{30}, or perhaps greater. In an infinitely small amount of time, the Universe reached the size of a football. This enormous expansion produced a drop in temperature. The Universe was supercooled from 10^{27} to 10^{22} K and then heated again in the reheating phase. Energy in the form of radiation gave rise to particle–antiparticle pairs that annihilated. Particles of matter and antimatter were created in pairs; they have the same mass but opposite charge, and if they meet, they annihilate, leaving only energy behind. However, today's Universe seems to be predominantly made up of matter. Where did all the antimatter go? The origin of matter is one of the greatest mysteries of physics. Given that the world is made of matter, it must be hypothesised that at some stage in the evolution of the Universe, an asymmetry between matter and antimatter was created. One might think that antimatter is segregated in some regions of the Universe, but very large regions of the sky have shown no evidence of radiation from border areas. As we will see, in 1967, the Russian physicist Sakharov published an article in which three conditions were introduced for the origin of the matter–antimatter asymmetry. Two of these conditions have been verified, but we are not sure about the third. Ultimately, to date, no mechanism has been found that can explain the problem of matter–antimatter asymmetry.

These events occurred between the grand unification era and an "extremely" long epoch, called the *electroweak era*, which developed in the period between 10^{-36} and 10^{-12} s. In this era, the Universe was full of particles and antiparticles that began to collide and therefore annihilate, leaving a surplus of matter particles that then formed our world. At that time, the energies were so high that photons collided to form particles of matter, which, however, when they encountered particles of antimatter, annihilated themselves – in short, a continuous flow of creation and destruction. At the end of the *electroweak era*, the *weak nuclear force* became a short-range force and separated into the electromagnetic force and the weak force. The Universe underwent another phase transition. This transition is referred to as the electroweak phase transition: the spontaneous breaking of the electroweak symmetry. The symmetry was broken by the *Higgs mechanism*, named after the famous Higgs boson, which provided mass to leptons (e.g. electrons, neutrinos), quarks, and other elementary particles. In the first billionth of a second (10^{-12} s), the Universe was made up of a plasma of quarks, leptons, and gluons, and its dimensions were approximately equal to the Earth–Sun distance. This quark soup, called *quark–gluon plasma*, was first observed at Brookhaven National Laboratory in 2002.

As time passes, the energy of the quarks decreased, and at around 10^{-6} s, when the temperature dropped below one thousand billion degrees, the Universe was no longer able to hold the quarks, which separated companions to form hadrons, a large family containing baryons (e.g. protons, neutrons). The baryons, in turn, contain an odd number of quarks and mesons, particles made up of two quarks.

Protons are almost eternal particles, while neutrons decay in 15 minutes. Meanwhile, the temperature continued to drop inexorably. The era just described is called the *quark era*. It extended between one billionth and one millionth of a second (10^{-12}–10^{-6} s).

The particles and antiparticles convert into photons, but the latter are no longer energetic enough to create particle–antiparticle pairs. The Universe was dominated by photons, electrons, neutrinos, and a minority (one part in a hundred million) of protons and neutrons. When the temperature dropped below 10^{12} K, this era, called the *hadronic era* (10^{-6}–1 s), consisting of a confinement of quarks to form hadrons, ended. We then arrived at the *leptonic era* (1 s–3 min), dominated by leptons and antileptons. Electrons and positrons collided and annihilated each other, generating photons, which in turn collided, forming other electron–positron pairs.

The decrease in temperature and density produced a decrease in reaction rates, with the consequent decoupling of neutrinos, when the Universe was at 1 s.

Returning to our history, the annihilation of most electrons had major effects on that of neutrons. In fact, in the *hadronic era*, neutrons did not disappear, as they were formed by the fusion of protons and electrons with the production of neutrinos. Following the annihilation of electrons, protons can no longer generate neutrons, whose number decreased. When the Universe was a little older than 1 s, for every 10 protons, there were only 2 neutrons. Neutrons are a little heavier than protons and decay in about 15 minutes. When the temperature was around 10^{10} degrees, neutrons were no longer created, and the neutron–proton ratio was frozen, as already mentioned, at around 1 neutron for every 10 protons. Due to the decay of neutrons, their number, at 300 s, when the temperature had dropped to around 10^9 degrees, was reduced to 2 against a number of 14 protons. The numbers of neutrons and protons were sufficient to form 1 helium-4 nucleus, leaving 12 protons. Since a helium nucleus is four times heavier than a hydrogen one, the fraction of helium-4 must have been given by 4 divided by $16(12 + 4)$; therefore, in the Universe, there were about 25% helium and 75% hydrogen. More in detail, when the temperature dropped below a few billion degrees, protons and neutrons gave rise to stable nuclei, such as deuterium, which do not have long lives. Things changed when the Universe was about 3 minutes old. Deuterium no longer decayed, and helium-4, helium-3, and lithium-7 were formed. Between 3 s and 20 min was the *era of nucleosynthesis*. As seen, the abundance of hydrogen (75%) and helium (25%) formed in nucleosynthesis depended on the density of protons and neutrons and radiation.

These numbers were confirmed by measuring the abundance of the elements formed and have remained unchanged to this day, despite the transformation of hydrogen into helium in stars. After about 20 minutes the temperature dropped to the point that nuclear fusion stopped. Primordial nucleosynthesis produced only light elements, up to beryllium. Since there were no stable nuclei with eight nucleons, nucleosynthesis stopped. The first calculations of nucleosynthesis were carried out in the 1940s (1948) by Ralph Alpher and George Gamow,[4] which were

---

[4] Gamow was a Ukrainian physicist, born in Odessa. With an eclectic character and a great sense of humour, he worked not only on cosmology and nuclear physics but also on molecular biology: he made a contribution to the deciphering of DNA. At 18, he went to

published in the famous article "αβγ", from the initials of Alpher, Bethe, and Gamow. Although the calculations had already been published in Alpher's thesis, Gamow thought of publishing an article in the journal *Physical Review*. Taken by his sense of humour, he added the name of Hans Bethe, for the reason that he himself explained thus:

> *it seemed to do a disservice to the Greek alphabet by signing the article with only the names of Alpher and Gamow, and so in preparing the report for the press the name of Dr. Hans Bethe was also included.*

Bethe accepted and contributed to the article, which appeared on April 1, 1948. Another fundamental prediction of Gamow, Alpher, and Herman was that the Universe must be immersed in a labile background of microwave radiation, the famous CMB, with a temperature of 5 K. Although this is not very well known, in 1950, Alpher and Herman managed to improve this estimate, obtaining the value of 2.8 K, a value very close to the one currently known (2.725 K). This prediction was forgotten until the 1960s, when Penzias and Wilson accidentally discovered the CMB.

At the turn of the era of nucleosynthesis, between 3 minutes and 240,000 years, we had the *era of radiation*. The Universe contained plasma, a glowing, opaque soup of protons and electrons. After the annihilation of leptons and anti-leptons, the energy of the Universe was dominated by photons that interacted with protons and electrons. Due to these continuous reactions, light underwent continuous deviations and reflections and was therefore trapped in the plasma. In the era between approximately 260,000 years and 380,000 years (temperature: 3000 K), the protons captured the electrons, forming neutral atoms. The phenomenon is known as recombination (of electrons and nuclei), associated with the decoupling between radiation and matter. Since the number of free electrons had decreased, the photons, trapped by their interactions with the electrons, became free to move and reach us. As the "fog" that enveloped the Universe disappeared, it became transparent, and a cosmic

---

study at the University of Novorossiysk and then in Leningrad (now St. Petersburg) with Aleksandr Friedman. Between 1928 and 1931, he studied in Copenhagen at the Bohr Institute for Theoretical Physics and at the Cavendish Laboratory. After a period in Göttingen, where he made a great contribution to nuclear physics, he was called back to Leningrad to work as a professor, but he definitively left Russia for America due to the Stalinist regime. He was also a great populariser of science.

background of visible light was released. This fossil background, due to the expansion of the Universe, is now observable in microwaves, which is precisely the CMB.

After the formation of atoms, practically nothing more happened for hundreds of thousands of years, apart from the fact that the Universe continued to expand and consequently cool.

The Cosmic Background Radiation

When the Universe expands, it cools, and the expansion "stretches" the wavelengths, increasing their length and decreasing their energy. The temperature decreases inversely proportional to the size of the Universe, and the wavelength of photons increases with the size of the Universe. Since energy is inversely proportional to wavelength, photons will see their energy halve as the size of the Universe doubles. The radiation coming from astronomical objects will be shifted towards the red part of the spectrum, i.e. it will be redshifted. This phenomenon is known as *cosmological redshift*. The expansion from the time of recombination to now has made the Universe magnify by about a factor of one thousand, similar to wavelengths. So, the photons of the background radiation are today a thousand times less energetic than at recombination. Similarly, the temperature from 3,000 K is reduced by a factor of 1,000; today, it has a temperature of 2.725 K and a wavelength of 7.35 cm and is observed in the microwave spectral region. As already mentioned, this radiation, which has the same temperature in all directions of the sky, is referred to as *Cosmic Background Radiation* (CMB), and is considered the fossil residue of the radiation emitted at the time of recombination and the big bang itself. This radiation was observed by chance in 1965 by two engineers: Arno Penzias and Robert Wilson, who were using a Bell telephone antenna to pick up the signals coming from the Echo-1 and Telstar satellites. They discovered microwave radiation that did not change with the time of day or orientation. Not having a preferential direction, it was not possible that it came from nearby New York or from the Sun. Initially, they thought it was produced by some interference or by the "white dielectric material", i.e. excrements of a pair of pigeons, as Penzias called them, which were housed in the antenna funnel. They consulted an astronomer from MIT, Bernie Burke, who knew about the studies of Robert Dicke and Jim Peebles at Princeton University. The scientists were planning an experiment to measure background radiation but were beaten by the unaware engineers. Penzias and Wilson published a short article in the *Astrophysical Journal* about what they had observed and won the Nobel Prize in 1978. In the previous pages of the same issue of the magazine, Dicke, Peebles, and collaborators explained the origin of the signal.

The discovery, together with the expansion of the Universe, and the abundance of light elements in the cosmos, was another confirmation and completion of the Big Bang theory. The radiation was evidence that the Universe had had a hot phase, confirming the ideas of de Sitter, Lemaitre, Gamow, and other scholars.

However, completely homogeneous and isotropic radiation would have implied the impossibility of the formation of cosmic structures and therefore also life. For this reason, physicists began to suspect that some inhomogeneities must exist. They were searched for and observed in 1992 by the COBE satellite in an experiment conducted by George Smooth and John Mather, who won the Nobel Prize for the discovery in 2006. The observations showed that the background radiation is isotropic up to 1 part in 100,000. In 1990, Mather, using COBE, also found that the radiation had the emission of a perfect black body, that is, an ideal object that absorbs all the incident radiation without reflecting it, with a temperature of 2.725 K. The observation confirmed Richard Tolman's prediction in 1934, in which he showed that blackbody radiation in an expanding Universe cools but maintains the same shape (it continues to be described by a blackbody distribution at different temperatures). So, the shape of the CMB spectrum, dating back to two months after the big bang, has remained unchanged to this day.

Given the great importance of the CMB, experiments were carried out to measure its characteristics with an increasing degree of precision. This was done with subsequent experiments on balloons, such as BOOMERANG, or satellites such as WMAP and PLANCK, which allowed an increasingly in-depth study of the inhomogeneities of the background radiation due to temperature differences and density fluctuations. Starting in 1997, experiments involving three flights of a high-altitude balloon, the BOOMERANG (an acronym for Balloon Observations of Millimetric Extragalactic Radiation and Geophysics), were carried out.

In 1997, the balloon flew in the skies of North America and in 1998 and 2003 in the skies of Antarctica. Like a real boomerang, taking advantage of the polar vortex, the balloon departed from the McMurdo Base, flew at an altitude of 42 km to reduce the absorption of the CMB

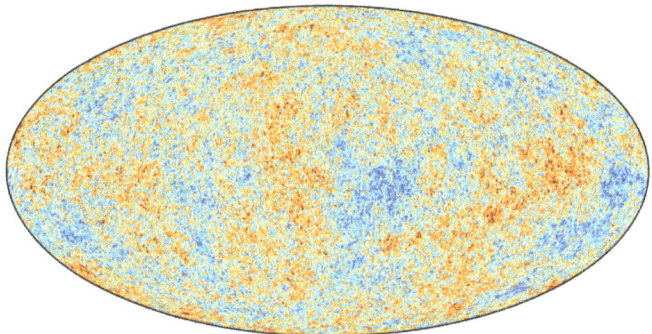

Figure 2.2.   CMB anisotropy map obtained from the PLANCK mission.
Source: ESA and PLANCK collaboration.

microwaves by the Earth's atmosphere, and with a closed orbit returned to the point of departure. The experiment directed by Paolo De Bernardis and Andrew Lange provided a high-resolution image of the anisotropies of the CMB in a portion of the sky. Given the success of the experiment, NASA planned a space mission, the Wilkinson Microwave Anisotropy Probe (WMAP), launched in 2001, which was followed by the European Space Agency's PLANCK mission, launched in 2009.

The satellites produced a map of the background radiation made up of warmer than average and colder areas, highlighting the temperature distribution of the plasma of which the Universe was made up 380,000 years after the big bang (Figure 2.2).

The map represents the small temperature variations coming from the surface where the photons last interacted with an electron, the so-called *surface of the last diffusion* (scattering), a sphere of the primordial Universe at the centre of which we are located, or rather the surface of the cloud from which the light was last scattered.

The fluctuations are isotropic up to 1 part in 100,000 and are visible on the map because they are significantly amplified. The hottest areas in the map correspond to the densest regions, so the map also represents density fluctuations in the early Universe, and could be defined as a photo of the Universe at the time of recombination. The scales and magnitudes of these fluctuations determine what our current Universe is. The fluctuations are loaded with information about the early Universe.

The Universe After Recombination

After recombination, the Universe was made up of hydrogen, helium, small amounts of deuterium formed during the time of nucleosynthesis, and hydrogen molecules (H_2). Due to the expansion of the Universe, the photons of the CMB lost energy, and the wavelengths were shifted to the infrared, causing the Universe to be deprived of visible light. Since there were no sources of light, such as stars, the Universe was transparent but dark. This period is called the *dark era* and developed in the period between 380,000 years and 300 million years after the big bang, when the first stars and galaxies were formed, which reionised the Universe, plunging it into darkness again, until about a billion years after the big bang. In the dark ages, the Universe remained dark until it was illuminated by the first stars. Stars formed from enormous clouds of gas, called *molecular clouds*, in gravitational collapse. Gravitational collapse manifests itself mostly in the densest regions of a molecular cloud and by the growth of the stellar embryo, which has a duration ranging from hundreds of thousands to millions of years. Larger stars form faster. During the collapse, the cloud becomes smaller and rotates faster and faster, forming an accretion disc, while the central temperature increases until nuclear reactions are triggered: a new star is born. After the star forms, the disc begins to cool, allowing grains of dust and ice to form. These begin to collide with each other, building increasingly larger objects which, influenced by gravity, form agglomerations. The objects grow up to a size of a few kilometres, which will then give rise, through aggregation, to *planetesimals*, the seeds from which the planets will form through growth. These stars, not yet observed, were different from the current ones. First of all, they were much larger than today's stars: a few hundred times larger than the Sun. The mechanism that produced the energy was, as in today's stars, nuclear fusion, and in particular, the dominant mechanism in the Sun: *proton–proton chains*, in which four hydrogen nuclei are transformed into one helium nucleus. During their lives, they produced elements heavier than hydrogen, and when they exploded as *pair-instability supernovae*, they scattered them into space. The daughters of these stars formed from the remains of primordial stars and contained, in addition to hydrogen and helium, traces of heavier elements (such as carbon). The primordial stars, called *population III*, released a significant amount of ultraviolet (UV) radiation over a hundred million years, which ionised the gas. The Universe became opaque again. The expansion of the Universe diluted the density of the plasma, until photons were once again free to move. So, the Universe experienced more than one *fiat lux*: with the big bang, after recombination, and finally at the end of reionisation. After the formation of stars, structures such as galaxies were formed. In the dark age, we are considering that, as in the rest of cosmic history, the expanding Universe contained a gas made of light elements and dark matter much more abundant than the gas. What happened to the gas in the Universe and to dark matter? In the early Universe, there were denser areas. These zones are the seeds from which structures in the Universe formed. Thanks to the concentration of mass, the force of gravity was able to overcome the expansion of the Universe, and the concentrations of dark matter were able to collapse to form the so-called dark matter halos. The first ones to form had masses of about a million suns. Gravity helped them grow, aggregating matter from their surroundings, so much so that today, the largest of these halos have masses of a million billion suns (10^{15} $M_\odot$).

Therefore, the dark matter first formed dark structures, halos, which, from a gravitational point of view, behaved like a sort of well into which the gas later fell, giving rise to visible objects, such as galaxies.

Matter–Antimatter Asymmetry

In the previous sections, we discussed how that "something" at the basis of our question "why there is something rather than nothing" originated. However, the problem remains that of understanding why, after that "something" was created, we did not end up with "nothing" due to the particle–antiparticle annihilations in the Universe. The big bang created an equal amount of matter and antimatter. Particles of matter and antimatter are created in pairs; they have the same mass but opposite charge, and if they meet, they annihilate, leaving only energy behind. However, today's Universe seems to be predominantly made up of matter. Where did all the antimatter go? The origin of matter is one of the greatest mysteries of physics. Given that the world is made of matter, it must be hypothesised that, at some stage in the evolution of the Universe, an asymmetry between matter and antimatter was created. With the use of an accelerator, such as the LHC, we are able to study the Universe up to about 10^{-12} s, when it was made up of a plasma of quarks and gluons. Before this moment, we can only rely on our theories and make extrapolations to energies never tested experimentally. It was in this time interval that the excess of matter over antimatter was generated, which gave rise to 4.9% of the mass-energy of the Universe. The remaining 95.1% is made up of dark matter and energy. One might think that antimatter is segregated in some regions of the Universe, but very large regions of the sky have shown no evidence of radiation from border areas. Today, there are two well-known paths to explain the matter–antimatter asymmetry. The first is the so-called *electroweak baryogenesis*.

In 1967, the Russian physicist Sakharov published the article "Violation of the CP Symmetry, C Asymmetry and Baryon Asymmetry of the Universe", in which three conditions were introduced for the origin of the matter–antimatter asymmetry:

1. The first condition is the violation of the C and CP symmetries. The first symmetry, the C, or charge conjugation, changes a particle into an antiparticle, while the P symmetry, where P stands for "parity", is the operation that changes the sign of the spatial coordinates, providing a mirror image of the physical system.
2. The second condition is the non-conservation of the baryon number B, i.e. in a reaction, the number of baryons is not conserved. This is a necessary but not sufficient condition because C would lead to opposite violations in the sign, which would cancel out any excess of the baryon number.
3. The final condition is that the reactions must occur outside thermodynamic equilibrium.

One may ask whether these conditions are met. In the 1950s, physicists Tsung Dao Lee and Chen Ning Yang proposed an experiment to verify the conservation of parity, P, in the weak interaction. In the winter of 1956–1957, Madam Wu and collaborators studied the decay of cobalt-60, showing that the P symmetry was not preserved. In 1964, Cronin and Fitch showed in an experiment with neutral kaons, K0, and their antiparticles that the CP symmetry was not respected. More recently, the Babar (in the Stanford Linear Accelerator Center in the USA) and Belle (in the High Accelerator Research Organization of Tsukuba, Japan) experiments have shown the violation of CP symmetry in several processes involving the so-called B mesons (including the b quark). It has been demonstrated that CP symmetry is not a symmetry of nature. As for the violation of the baryon number, it is possible to demonstrate its violation by considering times a millibillionth of a second (energy greater than

246 GeV) after the big bang. At times before a millibillionth of a second, the weak and electromagnetic interactions were unified in the electroweak interaction, where the mediators were massless particles (W_1, W_2, W_3, and the photon, γ). At a millibillionth of a second after the big bang, the *electroweak phase transition* occurred, in which the electroweak force split into weak and electromagnetic interactions, the first mediated by the massive bosons W^+, W^-, and Z^0, and the second by the photon. In 1985, Vadim Kuzmin, Valerij Rubakov, and Mikhail Shaposhnikov demonstrated that there were configurations, called *sphalerons*,[5] that allowed the transition between states with different baryon numbers and were not suppressed by increasing temperatures. These configurations gave rise to processes that violated the baryon number, maintaining thermal equilibrium. As expected, the breaking of the *electroweak symmetry* destroyed the thermal equilibrium of the sphaleronic processes, such that below the transition temperature, the violation of the baryon number was no longer observed. Therefore, the spontaneous breaking of electroweak symmetry gave rise to a scenario in which the remaining two conditions proposed by Sakharov were satisfied, namely the non-conservation of the baryon number B and the breaking of thermal equilibrium. To better understand the breakdown of thermal equilibrium in the electroweak phase transition, we can give an example. Let's consider a closed pot with a lid. Water molecules leave the liquid phase and pass to the vapour phase. When an equilibrium temperature is reached, the number of molecules that pass from the liquid phase to the gaseous phase is equal to those that perform the reverse process. If you remove the lid, the balance is broken. A similar thing happens in the *electroweak phase transition* (if of first order[6]), in which bubbles form in the electromagnetic phase and expand into the electroweak phase. The walls of the bubbles create the thermodynamic disequilibrium to satisfy the third Sakharov condition. Ultimately, we know that the CP symmetry is violated, the baryon number is not conserved due to sphaleronic processes, and the electroweak phase transition leads to the breakdown of thermal equilibrium. Therefore, in the electroweak transition, all the requirements set by Sakharov are qualitatively

---

[5] A *sphaleron* is a time-independent solution to the electroweak field equations and involved in processes that violate the baryon and leptonic number.

[6] First-order phase transitions are those that involve latent heat, such as that given off by ice when it melts.

verified.[7] The questions that can be asked at this point are how much baryon asymmetry is generated and whether it is sufficient to verify the second Sakharov criterion. In summary, although the mechanism of electroweak baryogenesis, within the standard model, qualitatively verifies the three requirements proposed by Sakharov for the generation of asymmetry, it does not work for two reasons: first, the CP symmetry violation rate is not sufficient, and second, the ineffective suppression of sphaleronic processes (which produce the violation of the baryon number) in the symmetry-breaking phase.

To solve the problem, the use of new physics, such as *supersymmetry*,[8] has been invoked (see Chapter 7). It was thought that CP symmetry violating supersymmetric sources would produce large baryon asymmetry. Unfortunately, the processes that produce this asymmetry generate intense *dipole moments*.[9] Failure to observe these dipole moments in elementary particles makes electroweak baryogenesis invoking supersymmetry infeasible. Another possibility is *GUT baryogenesis* at temperatures of ten million billion degrees. The problem is that the generated asymmetry could be eliminated by the electroweak phase transition. The other path that remains is that of *baryogenesis via leptogenesis*, i.e. the origin of baryons through that of leptons. Although the electroweak theory does not satisfy (quantitatively) the three Sakharov conditions using baryons, it could do so using other particles such as leptons (e.g. electrons, neutrinos). Sphaleronic processes involve leptons as well as baryons. For this

---

[7]More in detail, due to CP violation, the wall of the bubble has a selective permeability to the passage of quarks and antiquarks. If the flow of quarks towards the inside of the bubble is favoured, an excess of quarks will form inside the bubble and an excess of antiquarks outside. Meanwhile, the bubbles expand until they occupy all the fluid. The external antiquarks should fall into the bubbles, and upon meeting the quarks, cancel out the asymmetry. However, this does not happen due to sphaleronic processes. Outside the bubble, antiquarks dominate; processes that increase the number of quarks or reduce the number of antiquarks dominate the inverse processes until equilibrium is re-established and the asymmetry is cancelled. Inside the bubbles, the sphaleronic processes are much less intense than outside; therefore, we find ourselves with an excess of quarks. When the bubbles have occupied all the space and the electroweak transition is completed, we will be left with an excess of quarks.

[8] Supersymmetry is a theory that assumes that each boson corresponds to a fermion and vice versa.

[9] The dipole moment is a physical property related to a charge distribution that measures the separation of the centres of positive and negative charge.

reason, creating a leptonic asymmetry at high temperatures gives rise to a baryonic asymmetry. A model used is the one in which leptogenesis arises from the decay of massive neutrinos. There is a problem here too: it is not clear whether leptogenesis can be directly verified. The fact that neutrino masses are compatible with the demands of leptogenesis is encouraging, but it is not proof that the path is the right one.

Ultimately, to date, no mechanism has been found that can explain the problem of matter–antimatter asymmetry. This means that we are not able to answer the question "Why is there something instead of nothing?" with certainty.

Chapter 3

What the World Looks Like Seen
From a Ray of Light

*When a man sits with a pretty girl for an hour, it seems like a
minute. But let him sit on a hot stove for a minute – and it's
longer than any hour. That's relativity.*

— Albert Einstein

Relative to What?

Albert Einstein was a great revolutionary in physics, especially in his
youth. He gave a definitive form to special relativity,[1] something which
Henri Poincarè and Hendrik Antoon Lorentz, despite their notable contributions, failed to do. He participated in the demolition of classical
mechanics carried out by quantum mechanics. He overthrew Newton's
gravity with general relativity. The principle of relativity was not introduced by Einstein but was introduced into mechanics much earlier, in
1636, by Galileo Galilei. Galileo, in the *Dialogue on the Two Greatest
Systems of the World*, put these words into the mouth of Filippo Salviati:

> *Lock yourself up with some friends in the largest room under the deck of
> some large ship, and there keep flies, butterflies and similar flying*

---

[1] We talk about special relativity because this theory refers only to particular systems: those
that move with uniform motion, i.e. at a constant speed. We talk about general relativity
because this theory is an extension of the one restricted to non-inertial systems.

animals: also have a large vase of water, and some small fish in it; also suspend some bucket high up, which drop by drop pours water into another vessel with a narrow mouth placed below; and, keeping the ship still, carefully observe how those flying animals go with equal speed towards all parts of the room. [...] Observe that you will diligently carry out all these things, although there is no doubt that while the vessel is stationary they must not happen like this: make the vessel move with whatever speed you want; because (even if the motion is uniform and does not fluctuate from side to side) you will not recognize the slightest change in all the aforementioned effects; nor will you be able to understand from any of them whether the ship is moving or standing still.

Galileo highlighted that, confined within a laboratory (a ship in his case), there is no way to understand if you are stationary or if you are moving with uniform motion, i.e. with motion at a constant speed. This is the *principle of relativity* in mechanics, and in a nutshell, it says that the laws of physics have the same form for all inertial observers, i.e. observers at rest or in uniform motion. Said in more technical terms, physical laws must be invariant (not change) when passing from one inertial reference system to another. Let's explain a little better what we mean. Each observer of physical reality applies physical laws, starting with his own reference system. The reference system is made up of a point with respect to which distances are measured, a ruler with which we measure distances, and a clock with which we measure times. To simplify, considering the case of a ship: an observer on the beach constitutes a reference system and one on the ship another reference system. Now, although the quantities measured by one or another reference system may be different, physical laws cannot depend on the observer. This is one of the fundamental principles of modern physics: physical laws must be the same in different references, and the basic aim of relativity is to find an invariant form for the laws of physics.

However, with the introduction of electromagnetism, problems arose. It seemed that the principle of relativity did not apply to the laws of electromagnetism and that the latter changed shape depending on the reference system considered. In particular, the *Galilean transformation* of speeds was not applicable to electromagnetic waves. What does this transformation consist of? It is a simple speed composition rule; it states that when passing between two different reference systems, the speeds must be added vectorially. Let's take an example. Suppose we have a car travelling at 50 km/h and a pedestrian standing on the pavement. In the car's reference system, i.e. for the driver inside the car, the driver's speed with respect to the passenger compartment is 0 km/h. However, for the pedestrian standing on the pavement, i.e. having a speed of 0 km/h, the car is moving at a speed of 0 km/h + 50 km/h, i.e. 50 km/h, the sum of the speed of the pedestrian plus that of the car. If the pedestrian moves on the pavement at a speed of 5 km/h in a direction parallel to that of the car, then the speed of the car for the pedestrian will be 50 km/h − 5 km/h, i.e. 45 km/h . Finally, if the pedestrian moves at 5 km/h parallel to the car but in the opposite direction, the speed of the car for the pedestrian will be 50 km/h + 5 km/h, i.e. 55 km/h. This is the speed composition rule in the Galilean transformation of speeds. As mentioned, using this rule, the laws of electromagnetism changed as the reference changed. There were two solutions to the problem. The first solution was that the laws of electro-magnetism were wrong, but this was not possible because the laws were in agreement with experimental evidence. The second solution was that there existed a privileged reference system in which the laws of electro-magnetism took their known form. There were physicists who followed the second path and introduced the concept of *luminiferous ether*, assum-ing that the favoured system was "ether", which was also the medium in which light waves propagated. However, it was immediately clear that this idea was not the right one. First of all, it had to possess strange physi-cochemical characteristics: it had to be solid and rigid to support the high speed of light; it had to be immobile and, at the same time, not offer resis-tance to the motion of celestial bodies. Experiments were carried out to verify the existence of the ether. The most famous is that of Michelson and Morley in 1887.

The first experiment was carried out in 1881 by Albert Abraham Michelson using an interferometer (Figure 3.1). Suppose that the ether is integral with the Sun. The Earth, in its revolutionary motion, moves with respect to the ether with a speed of v. Even the interferometer, situated on Earth, moves with speed v with respect to the ether. The speed of light with respect to the ether is $c = 299,792,458$ m/s. The interferometer works in the following way. Light from a source is directed towards a semi-sil-vered mirror, P, which splits the beam. Half of the light is reflected and reaches the mirror S_2, where it is reflected again and reaches the screen I. The other half passes through P and is first reflected by the mirror S_1 and then by P, after which it reaches the screen. The two beams form an inter-ference pattern on the screen, i.e. bright areas alternating with dark areas.

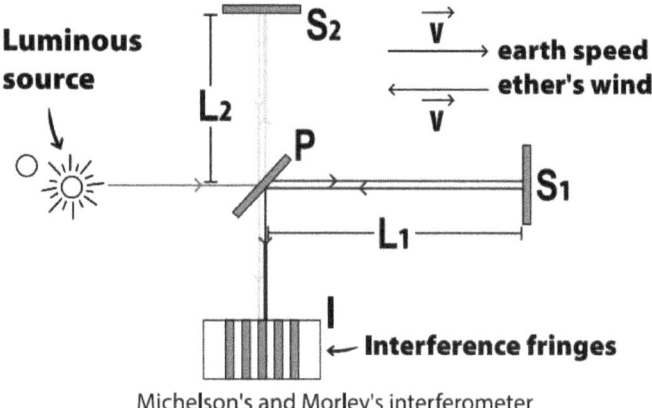

Michelson's and Morley's interferometer

Figure 3.1.   Michelson–Morley interferometer.

Source: Reproduced from https://www.youmath.it/lezioni/fisica/teoria-della-relativita-ristretta/3579-esperimento-michelson-morley.html

Pointing the interferometer in the direction of the Earth's velocity and then at 90° should have produced a sliding of the interference fringes, which was not observed. The same result was obtained in Michelson's 1887 experiment, carried out with Morley using an improved device. The result indicated that the ether did not exist and that the speed of light did not depend on the reference frame. It was the same whether you approached a beam of light, at any speed, or moved away from it.

In 1905, Einstein published an article entitled "On the Electrodynamics of Moving Bodies", i.e. the first formulation of the theory of special relativity. The article began with two postulates. The first was the principle of relativity extended to all physics. The second was that the speed of light in a vacuum is constant and that it does not change as a function of the state of motion of the source or the detector. The conclusion Einstein reached was that Galileo's transformations were incorrect, and he determined new transformations, which coincided with the *Lorenz transformations*, derived by the physicist Lorenz in 1904, to ensure that the laws of electromagnetism were invariant. The implications of the new transformations went against common sense: they implied a slowing down of time and a contraction of distances as speed increased, but they solved the problem with electromagnetic forces and "explained" why light moved at the same speed for all observers. Special relativity removed the states of

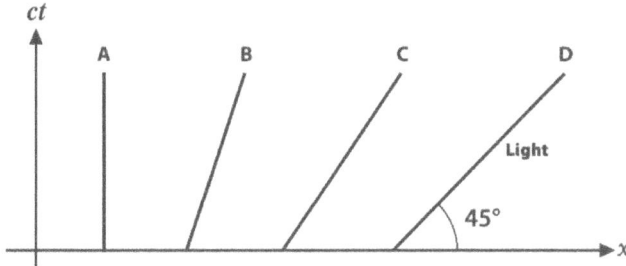

Figure 3.2.   Worldlines of particles moving at constant speed.

absoluteness and eternity from space and time. Space and time became concepts relative to the observer and were no longer inert and immutable quantities. They merged into a single space-time entity, introduced in 1907 by a Polish-German mathematician, Herman Minkowski, who had been Einstein's teacher. Space-time is a dynamic entity; it can dilate, shrink, and curve like a piece of rubber. All phenomena take place in space-time. A point in space-time is called an event and is represented by three spatial coordinates and one temporal coordinate. The transition from one reference system to another is given by a rotation in the four-dimensional space made up of the three positions and time, which keeps the space-time distance between two events unchanged. The history of a particle is represented by a curve called the *particle's worldline*. To give an example, let's draw a Minkowski diagram, shown in Figure 3.2.

Space-time is four-dimensional, but we cannot draw it. Let's consider a two-dimensional section: the *x*-axis represents space, and *ct* represents time. The vertical line, A, represents a particle at rest. The lines *B* and *C* move, the latter with a speed greater than that of the former. A ray of light in this diagram is, by convention, indicated by a line inclined at 45°. Let's go back for a moment to counting distances and time dilation to see how such effects can be explained with thought experiments.

It is said that Einstein came up with the idea that time can expand on a beautiful Sunday while walking with his wife and son in a park in Bern. Sitting on a bench, he remembered a dream he had as a teenager. In the dream, riding a beam of light, you start with a clock that shows noon. In Einstein's dream, time had stopped, and the clock always showed noon. To understand time dilation, you can do a similar thought experiment. Let's consider a light clock. This clock consists of two mirrors

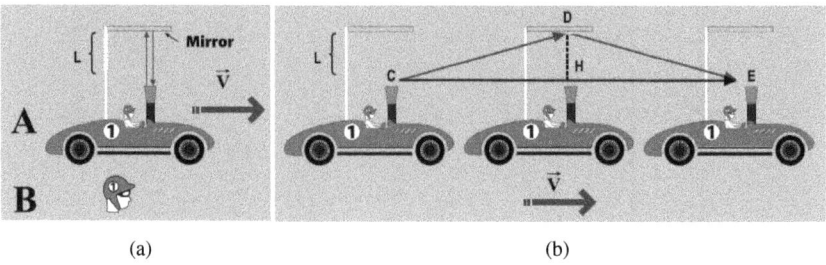

Figure 3.3.   Time dilation.

Source: Reproduced from https://digilander.libero.it/antonio1001/relativita/dilatazione.htm.

(or similarly, as in Figure 3.3, a flashlight that emits light onto a mirror), and a ray of light bounces between them. The device behaves like a clock: every time the ray touches the upper mirror, the clock ticks. Now let's see how the situation appears to an observer, *B*, who is stationary, and another, *A*, who is moving in a car. At a certain moment, observer A sends off a short light signal. Observer *B* stands still and observes. What does observer *A* see? For him, the ray of light moves towards the mirror and then returns back, covering a distance of 2*L*. To travel the distance of 2*L*, the light will take a time interval Δ*t* equal to 2*L* divided by the speed of light. What does observer *B* see? For him, since the car moves at a speed of *v*, he will see the ray of light start at point *C*, touch the mirror at *D*, and then arrive at the torch at *E*. That is, the path of the light will no longer be 2*L* but greater. Since the speed of light is constant and time is proportional to the distances travelled, for observer *B*, the measured time interval is greater than for observer *A*, i.e. $\Delta t_B > \Delta t_A$; for the moving observer, *A*, time passes more slowly. To find the relationship between the two intervals, simply use the Pythagorean theorem.[2] However, there is a little problem. If you look at things from A's point of view, it is *B* who is on the move; therefore, time should pass more slowly for *B*. Which of the two observers

---

[2] Using the Pythagorean theorem, we can write $CD^2 = CH^2 + L^2$. We can also write $CD = c\Delta t_B/2$, $CH = v\Delta t_B/2$, and $L = c\Delta t_A/2$. Substituting these quantities into the Pythagorean theorem, we can find the relationship between the times for observers *A* and *B*: $\Delta t_B = \Delta t_A/\sqrt{(1 - v^2/c^2)} = \gamma\Delta t_A$. The term $\gamma = 1/\sqrt{(1 - v^2/c^2)}$ is larger than 1; therefore, $\Delta t_B > \Delta t_A$, and it is concluded that for the moving observer, A, time passes more slowly. For example, moving at 99% of the speed of light, 100 years for the stationary observer correspond to approximately 14 years for the moving one.

is right? Both – in their own reference systems. We are in a paradoxical situation. We can resolve this apparent contradiction by observing that if the time interval refers to two events, then all observers agree in saying that the observer who sees the events occur in the same place also measures the shorter time. In fact, observer A sees the events "departure of the light" and "arrival of the light" occur in the same place (on the flashlight), while observer B sees the two events occur in different places. Therefore, for observer A in motion, time passes more slowly. Another way of resolving the paradox is to note that when observer B wants to measure A's time, it is necessary for their clocks to be synchronised since observer B must make two measurements in two distinct places as observer A is moving; therefore, B must use two clocks. The fact that B uses two clocks introduces a difference in the ways of measuring time. The paradox begins here since two events that are simultaneous in one reference system are not simultaneous in the other. Two clocks placed in different places will show the same time in a system at rest but different times in a system in motion. Therefore, for A, B started measuring time at different times, which explains the paradox. The time measured by the moving observer, i.e. the time along the side DH of the triangle CDH, is called *proper time*. Looking at the triangle CDH, we see that, according to the Pythagorean theorem, the square of the side DH is given by the difference of the squares of CD and DH; therefore, the proper time is always less than the time for someone who stands still.[3]

You can also show that the lengths contract along the direction of motion with another thought experiment (Figure 3.4). Let us again consider the two observers, A and B. Both observers want to measure the length of the car.

Observer A can measure the length of the car by taking, with his watch, the time Δt_A elapsed between two event indicated by the red arrows in Figure 3.4 (the passage of the tip of the car from a fixed point and the passage of the tail of the car from a fixed point). Since for A, the length of the car is proportional to Δt_A and for B to Δt_B and since the time for A is dilated, it will result that $L_A > L_B$, i.e. the length of an object, along the

---

[3]If we call τ the proper time, we can write using the Pythagorean theorem that $DH^2 = (c\tau)^2 = CD^2 - CH^2 = (c\Delta t_B)^2 - (v\Delta t_B)^2 = (c\Delta t_B)^2 - x^2$. That is, the proper time of the moving observer is $\tau^2 = \Delta t_B^2 - x^2$, always lower than the time for someone standing still because the distance traveled by the clock is subtracted from the observer's time.

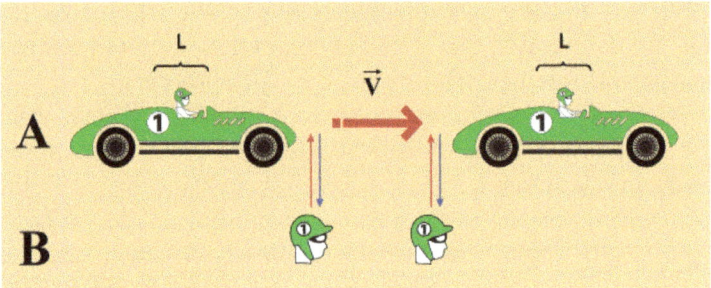

Figure 3.4.   Contraction of lengths.

Source: Reproduced from https://digilander.libero.it/antonio1001/relativita/dilatazione.htm.

direction of motion, appears to be shorter for an observer who is not at rest with respect to the object.[4]

Faster than Light?

A fundamental point to remember is that light, in addition to having the same value regardless of the state of motion of the source and the detector, is the maximum speed that can be reached in a vacuum. This might seem absurd because, by maintaining, for example, an acceleration equal to that of gravity, 9.81 m/s^2, for more or less a year, we will reach the speed of light. There is a problem, however. One of the results of special relativity is that mass and energy are essentially the same thing, as the very famous formula $E = mc^2$ states. When we accelerate an object, the energy we impart on it goes into increasing its mass to a very small extent. As the speed increases, more and more energy must be supplied to further increase the speed because more and more energy is transformed into mass. The closer we get to the speed of light, the more massive and immovable the object becomes. Moving at 99.9% of the speed of light, an

---

[4]For A, the length of the car is $L_A = v\Delta t_A$. Observer B measures in the same way by taking with his watch the time Δt_A elapsed between the two events marked by the blue arrows. For B, the length of the car will be $L_B = v\Delta t_B$. Dividing the two relations to compare the measurements, we obtain $L_B/L_A = \Delta t_B/\Delta t_A$; therefore, $L_A = \gamma L_B$, and since $\gamma > 1$, the result will be $L_A > L_B$. For example, a 100 m spaceship moving at 99% the speed of light will appear to be approximately 14 m long.

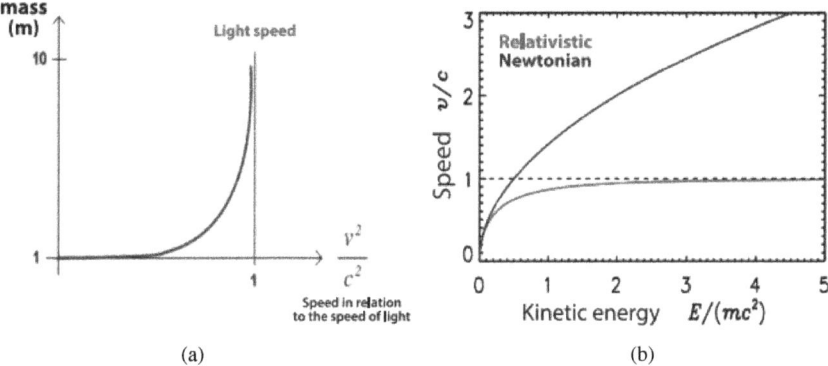

Figure 3.5.   (a) Relativistic mass and (b) kinetic energy.

80 kg man would have a mass of approximately 1,800 kg. In Figure 3.5(a), it is shown how the mass changes approaching the speed of light, *c*.

As can be seen from the figure, the closer we get to the speed of light, the faster the mass increases, while the speed increases very little. In Figure 3.5(b), we see how by increasing the kinetic energy of an object, i.e. the energy due to its motion, the speed grows less and less as the energy increases. To accelerate a body to the speed of light, an infinite amount of energy would be needed, and for this reason, it is impossible for an object with mass to reach it. To give some examples, suppose we have a very light spacecraft, weighing only 10 tons. If we try to accelerate the spacecraft beyond 70% of the speed of light, we would need an energy of 1.26×10^{21} J. To get an idea of the magnitude of this energy, let's compare it to that produced by the European Union. In 2022, the European Union produced 2,641 TWh,[5] i.e. 9.51×10^{18} J. So, the energy needed to accelerate the spacecraft to 70% of the speed of light is that produced by the European Union in 133 years! To bring it to 99% of the speed of light, it would require a fivefold increase in energy. If we tried to push the spacecraft to a speed ever closer to that of light, the energy needed would grow enormously, until it becomes infinite. Only objects that have no mass can reach the speed of light, such as photons, the particles of light. So, is there no way to exceed the speed of light? First of all, it must be pointed out that only in a vacuum is the speed of light the maximum speed in our Universe.

---

[5] TWh, i.e. tera watt hour, where tera equals 10^{12} and Watt is the unit of measurement of power and is equal to one Joule per second.

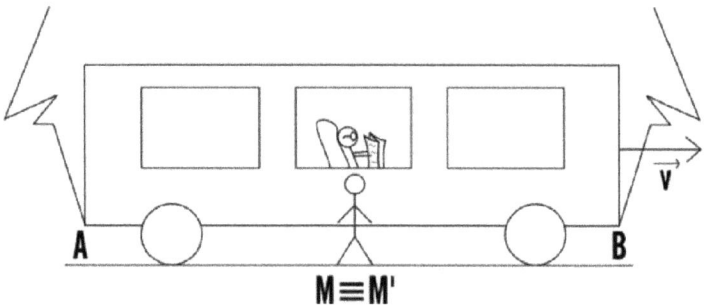

Figure 3.6.   Relativity of simultaneity.

If we consider any other medium, such as air or water, the speed of light is less than its speed in a vacuum. For example, in water, the speed of light drops to around 226,000 km/h. In a medium, it is possible to exceed the speed of light in that medium. For example, a particle such as a muon can exceed the speed of light in water. When this happens, we observe the emission of waves in a cone, similar to what happens in the air when, for example, an aircraft overcomes the "sound barrier". This is called the *Cherenkov effect*. A Danish physicist and a professor at Harvard, Lene Hau, managed to slow the speed of light to 61 km/h in 1999 and then, in 2001, managed to stop it. Lene Hau and her team sent a beam of light towards sodium cooled to a few millionths of a degree above absolute zero. The body at these temperatures behaves like a single atom. The light beam was first absorbed and then released after being hit with a laser beam.

Finally, although special relativity says that nothing can travel faster than light, the theory still does not prohibit space itself from expanding at superluminal speeds. Another important point is that, due to the relativity of time in special relativity, the simultaneity between two events becomes relative to the observer. To explain this, we use a famous thought experiment proposed by Einstein (see Figure 3.6).

Consider a train moving with speed v to the right. Consider an observer, S', standing inside the train, and another, S, standing on the platform. Two lightning bolts simultaneously strike the train at points A and B. Suppose that when this happens, the observer on the platform is at the midpoint, M, between A and B, and the observer on the train is also at the midpoint, M', between A and B. For the observer on the platform, the light from the two lightning bolts will arrive at the same time, being at the midpoint; therefore, for him, the two lightning bolts have fallen simultaneously. What happens to the observer on the train? Since he moves to the right with speed v, he will first see the lightning strike B and then A. For him, the two events (the fall of the two lightning bolts) are not simultaneous. Therefore, simultaneity becomes relative: it depends on the observer and the state of motion.

What Evidence Do We Have That This is the Case?

Thinking back to what was said earlier, a whole series of things happen in special relativity that we do not observe in everyday life. Lengths shorten in the direction of motion, time expands, simultaneity is relative to the observer, and so on. A natural question is whether what special relativity predicts is true. Well, a whole series of experiments have been carried out to verify the predictions of the theory, and all of them have shown that the predictions are correct. A first experimental confirmation was obtained by Bruno Rossi and David B. Hall in 1940 regarding the average life, i.e. the average time that must pass before it decays, of some particles such as pions and muons. Let's review the experiment. A muon is a particle of the electron family but with approximately 206 times greater mass. This particle is unstable and decays. Its average life is equal to 2.2 millionths of a second (2.20×10^{-6} s) when stationary. Particles (mainly protons) called primary cosmic rays arrive from space. When they enter the atmosphere, they interact with its molecules, forming other particles called secondary cosmic rays, which move towards the ground. Every second, every square centimetre of the Earth's surface is struck by a particle. Muons are part of secondary cosmic rays. If a muon originates at a height of 4 km and moves at a speed of 99% of the speed of light, it will take 1.35×10^{-5} s to travel 4 km, a time longer than its decay time. Therefore, the muon should not be able to reach the ground. Yet, muons arrive on Earth's soil. How? We have seen that when stationary, muons decay in 2.2 millionths of a second. If we calculate the average life time when the muon moves at 99% of the speed of light, we find that the average life time is 1.56×10^{-5} s, i.e. for an observer stationary on Earth, the muon lives approximately seven times longer than at rest, enough time for the muon to arrive on the Earth's surface.

We can think in another way and see what 4 km corresponds to for the muon due to the contraction of the lengths. Using special relativity, for the muon, 4 km corresponds to 564 m, and to travel this distance in its reference system, the time needed is 1.9×10^{-6} s, and the muon arrives on the Earth's surface. Whether looking at things from the point of view of the observer on Earth or from that of the muon, the result is the same. The dilation of time and the contraction of distances explain the so-called "muon mystery". Another experiment is the one conducted by Joseph Hafele and Richard Keating in 1971 to test both special and general relativity, and we will talk more about it in the following chapter. Time dilation and distance contraction are continuously observed in particle accelerators. In 2022, the contraction of the electric field around an electron moving at speeds close to that of light generated by an accelerator was observed. Special (and also general) relativity is confirmed by the global positioning system (GPS), which we use to orient ourselves on Earth. The system uses satellites that move at speeds of the order of 3,800 m/s. This produces a time dilation of 1.00000000008, and in one day (86,400 s), the clock in the satellite slows down by 00000000008 × 86,400 = 7 millionths of a second. Multiplying by the speed of light, we find that after one day, we would have an error of 2,100 m, making the GPS unusable. For this reason, this correction must be taken into account together with another due to general relativity, which we will see later. Relativity also affects our daily lives.

Chapter 4

Does Mass Bend Space-Time?

Unthinking respect for authority is the greatest enemy of truth.

— Albert Einstein

The Most Beautiful of Theories

Despite the revolution produced by special relativity, the space-time defined by it continued to have characteristics of classical absolutism: it was different for each observer but could not be modified by the observer. Furthermore, special relativity did not consider a large class of motions; it referred to systems with uniform motion. Einstein considered generalising the theory to systems in non-uniform motion, creating what is now known as the *theory of general relativity*. The first steps that led him to this theory were made with ideal experiments and the formulation of the so-called *equivalence principle*, which states that around any point, it is always possible to find a reference system in which the effects of acceleration due to the gravitational field are zero. Using the equivalence principle, Einstein extended his theory of special relativity to non-inertial systems. Einstein used this principle as a guide in his search for the equations of general relativity, the field equations. The theory was presented in four communications to the Prussian Academy of Sciences in November 1915, and he explained it fully in an article in the *Annalen der Physik* of May 1916. The meaning of Einstein's field equations can be expressed as follows:

Curvature of space time = Density of matter and energy

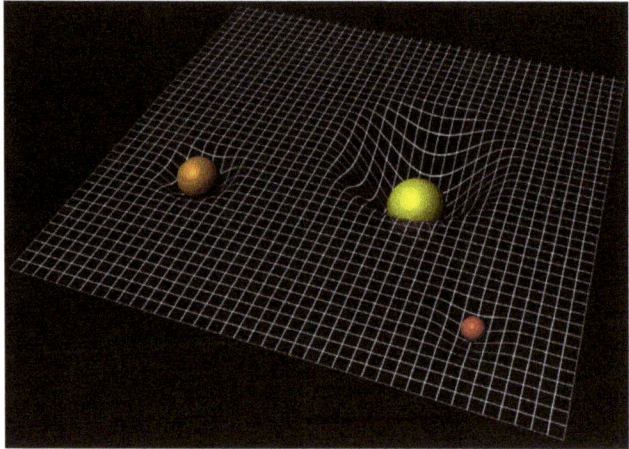

Figure 4.1.   Masses that deform space-time.
Source: ESA-C. Carreau.

In other words, the field equations, 10 equations in four-dimensional space-time, relate the curvature of space-time and the sources of the gravitational field (matter and energy). Similar to the theory of special relativity, in general relativity – "the most astonishing combination of philosophical insight, physical intuition and mathematical skill" in the words of Max Born and "the most beautiful of scientific theories" in those of Lev Landau – space and time are not two separate and immutable enti-ties but are united in marriage. They form four-dimensional space-time, made up of three spatial coordinates and one temporal coordinate, which behaves like a plastic structure. Both space and time can contract, expand, lengthen, or shorten. All this depends on mass. Unlike what happens in special relativity, space-time is not simply relative to the observer; the observer actually modifies the geometry of space-time. Time flows slower in strong gravitational fields, as we will see in the following. Therefore, for a person living at sea level, time passes more slowly than for another person living on the Everest since the gravitational field is stronger when we approach the centre of the Earth. The time difference, summed over an entire human lifetime, is a few billionths of a second. In a well-known analogy, space-time is described as a stretched rubber sheet. An object placed on the sheet will deform it, producing a depres-sion (Figure 4.1) and changing its curvature.

Figure 4.2.   Motion of a planet around a star due to the deformation of space-time.

Gravity is no longer a force as in the case of Newtonian mechanics but is generated by the curvature of space-time produced by massive objects (Figure 4.2). In the presence of a mass, space-time deforms, like the surface of a rubber sheet. Gravity consists of this space-time deformation that forces objects to move on a curvilinear trajectory. The trajectories of the bodies are curves called *geodesics*, which are the curves of minimum distance in curved space-time. For example, on the Earth's surface, geodesic lines are portions of great circles similar to the equator.

Moving away from the analogy, the best way to express what happens and a good way to summarise the theory is the following quote by John Wheeler: "matter tells space how to curve, space tells matter how to move". Given its geometric nature, it is therefore natural to hypothesise that gravity, i.e. the space-time deformation, also influences the propagation of light. If we have a luminous object that emits a ray of light directed towards an observer, and if there is a body interposed between the body emitting the light and the observer, the light ray will change its trajectory due to the space-time deformation.

Falling Graves

To see the most important predictions of the theory, let's go back to the *equivalence principle*. The concept of mass was introduced with Newton's second law, which says that force is given by the product of acceleration and a characteristic of the body, which is precisely the mass. Mass is a quantity that measures inertia, which indicates the tendency of a body to

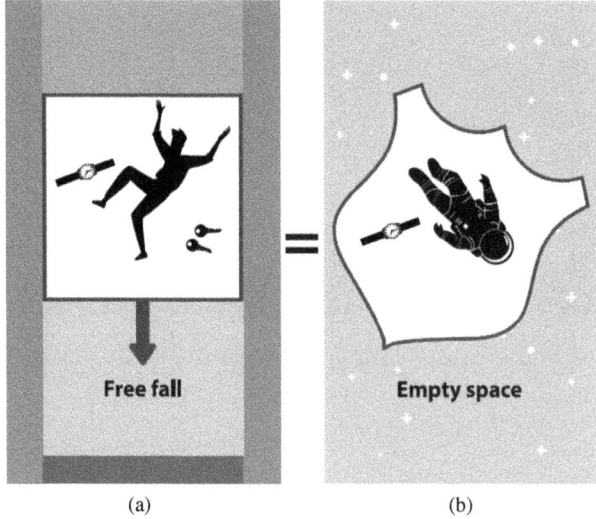

Figure 4.3.   Both a person in a lift in free fall, 4.3a, and an astronaut in orbit, 4.3b, float in the air, weightless.

oppose movement. We therefore speak of *inertial mass*. Newton also intro-duced the law of universal gravitation, which says that two bodies attract each other in a manner proportional to the mass of the two bodies and inversely proportional to the square of the distance. The mass that produces the attraction is called *gravitational mass*. A series of experiments have shown that the two masses are practically the same, and this result is called the *weak equivalence principle*. The fact that gravitational and inertial masses are the same thing implies that all objects experience the same gravitational acceleration. On the other hand, the fact that all massive objects fall with the same acceleration has been well known since Galileo's time. Thinking about this aspect of gravity allowed Einstein to extend his theory of special relativity. Why are the two masses the same? Einstein performed a famous thought experiment, that of an elevator. Suppose we are in a closed elevator from which we cannot see what is happening out-side. At a certain moment, the cable breaks, and the elevator falls. What will happen to the people inside? They would begin to float in the air (Figure 4.3(a)), the same thing that happens to astronauts in orbit (Figure 4.3(b)).

Now, suppose we take the elevator into space, away from any object. We would find ourselves floating in the air, like when the elevator was falling. Now, suppose you put a rocket engine at the base of the elevator in space and turn it on.

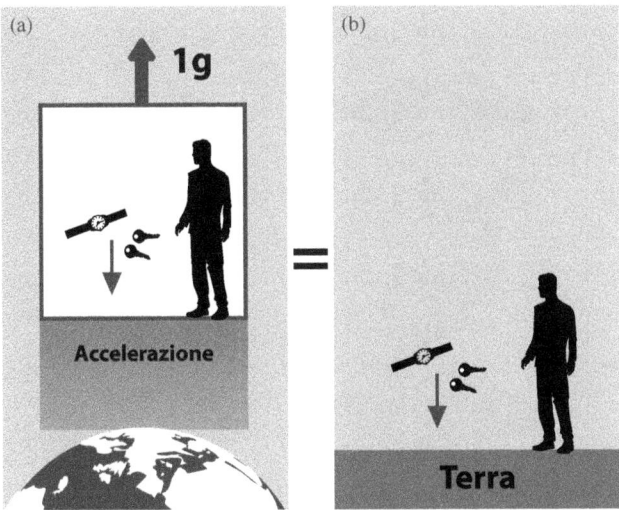

Figure 4.4. Gravity is equivalent to an accelerated laboratory: (a) the laboratory accelerating in space and (b) the laboratory on Earth.

The elevator accelerates, but inertia opposes the motion, and we would fall onto the base of the elevator (Figure 4.4(a)). The situation is similar to that of the elevator on Earth (Figure 4.4(b)): you would feel attracted downwards as if there were a gravitational field. If we asked the occupant of the elevator whether he was in space or on Earth, he would not be able to answer because he has no way of distinguishing between the effects of gravity and those produced by acceleration.

Therefore, the mass of a body attracted towards the Earth (the gravitational mass) cannot be different from that of a body subjected to an acceleration equal to the Earth's gravitational acceleration in the absence of gravity (inertial mass). This aspect of acceleration and gravity posed a problem for the theory of special relativity, for which reference systems must move at a constant speed. Since gravity is equivalent to an acceleration, it is not possible to define an inertial reference system in the presence of gravity. We had to find a way to eliminate gravity. It was then, in November 1907, that Einstein had an idea, as he wrote, "the happiest idea of my life":

I was sitting on a chair in the patent office in Bern when suddenly a thought came to me: "If a person falls freely he will not feel his own weight." I was surprised. This simple thought made a deep impression on me. It pushed me towards a theory of gravitation.

Figure 4.5.   Gravity abolished in a free-falling laboratory: (a) free-falling laboratory and (b) laboratory floating in space.

To eliminate gravity, you just need to surrender to it and let yourself fall (Figure 4.5), so it was possible to continue using the ideas of special relativity, at least locally, in a not-too-large region of space.

Einstein took these conclusions as the central point of his theory, which was called the *equivalence principle*. According to this principle, there is no way to distinguish between acceleration and a gravitational field, and all non-rotating reference systems in free fall are equivalent when performing physical experiments. In other words:

> *For any frame of reference in free fall within a gravitational field, the laws of physics are the same as those that apply to an inertial frame, without gravity.*

The equivalence principle holds only in small regions of space-time, so that the curvature is not detectable and the space-time is flat. By analogy, if we observe the surface of the Earth in a small region, we will not notice that it has a curvature and that it is more or less spherical, but it will appear flat. In such a region, the gravitational field is constant, and objects move with the same acceleration and in parallel lines. In a large region of space-time where curvature effects are present, free-falling observers will feel the curvature effects in the form of *tidal forces*. The latter are gravitational forces of compression and stretching, discovered by Isaac Newton, and on Earth, for example, they cause the tides.

Einstein continued to carry out mental experiments, starting with the principle of equivalence, and arrived at some surprising conclusions. Let's see some of them.

Masses Curve Space-Time

Suppose we are inside a container suspended above the ground by a cable (Figure 4.6). A photon leaves the left wall, and at that moment, the cable is cut. As a result, gravity has been eliminated, and the container behaves like an inertial system. An observer inside the container, according to the equivalence principle and the laws of physics, would see the photon move to the right in a horizontal manner. An outside observer would see that the container is falling due to gravity. The external observer would see that the photon follows a curved trajectory pointing downwards.

Why? The light continues to move along the fastest path. Before cutting the cable, this trajectory is a straight line, but after that, it is a curve. It is not the light that has changed direction, but the space that has curved. The curved trajectory of the photon is the shortest trajectory in curved space-time around the Earth. This experiment shows two key points. The first is that space curves in the presence of gravitational fields (Figure 4.2), and the second is that light follows the curvature of space. So, if a ray of light passes near a massive object, such as the Sun, it is bent (see Figure 4.7).

This effect can be used to test Einstein's theory, as was done in 1919, as we describe in the following. Another effect related to the latter is *gravitational lensing*. Due to the lensing effect, images of distant galaxies are distorted by the formation of multiple images of an astronomical object, or rings, due to the presence of an object between the observer and the source (see Figure 4.8).

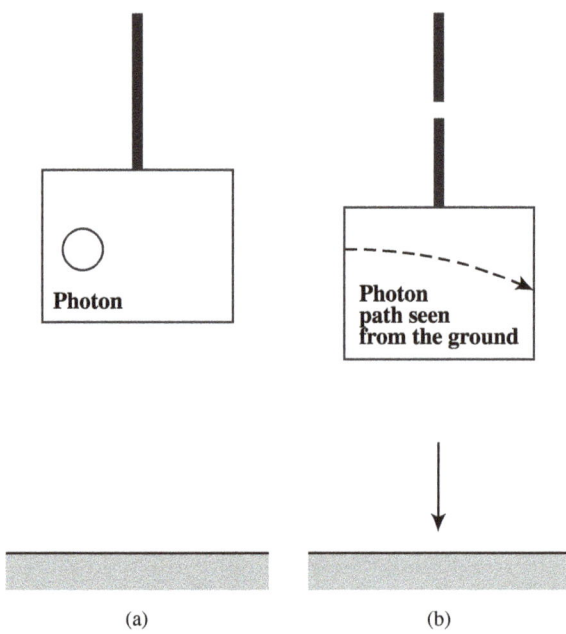

Figure 4.6. Equivalence principle for a photon moving horizontally: (a) the photon leaves the left wall and (b) arrives at the right wall.

Source: Adapted from Carrol & Ostlie (2017). *An Introduction to Modern Astrophysics*. Cambridge University Press.

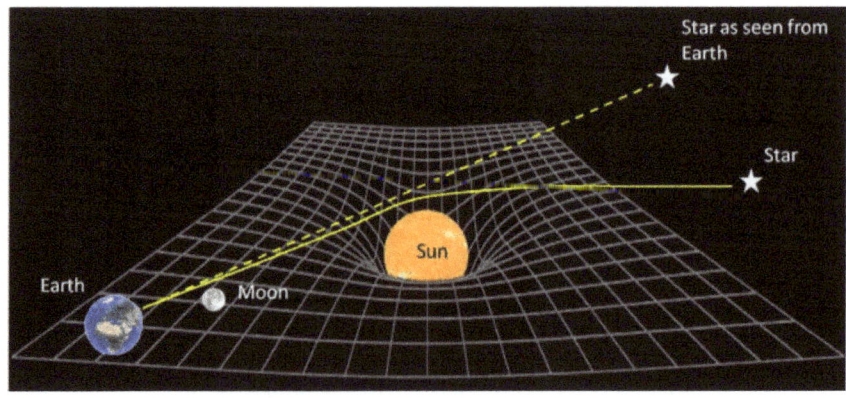

Figure 4.7. Deflection of light rays from the Sun.

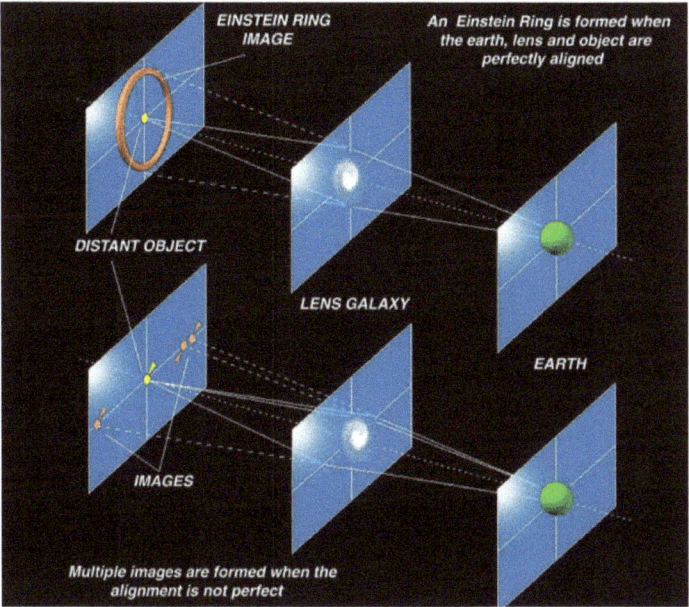

Figure 4.8. Formation of a ring and multiple images. An Einstein ring forms when the alignment between the Earth, the galaxy creating the lens, and the distant galaxy is perfect.

Source: University of Manchester/Alastair Gunn.

Gravity and the Passage of Time

We continue to follow Einstein in his thought experiments. We again consider our closed container suspended by a cable (Figure 4.9). A photon of a certain frequency leaves the base of the container, and simultaneously, the cable is cut. When the container falls, gravity is eliminated. Due to the equivalence principle, the container is an inertial system, and when the photon reaches the ceiling, it must have the same frequency. For an observer outside, the container is falling, and the photon moves upwards, "fighting" against the gravitational force. When the photon reaches the top of the container, it will have undergone a change in frequency; the frequency will have shifted towards the red. This effect is called *gravitational redshift*.

Let's try to intuitively understand why this happens. If we throw a stone upwards, it will fight against the Earth's gravity and will tend to lose

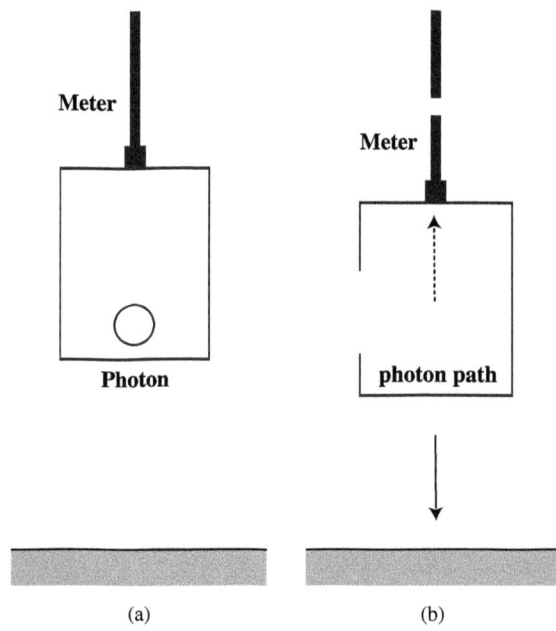

Figure 4.9.   Application of the equivalence principle to light propagating vertically. The photon (a) starts from the floor and (b) reaches the roof.

Source: Adapted from Carrol & Ostlie (2017). *An Introduction to Modern Astrophysics.* Cambridge University Press.

energy and slow down. Light always moves at the same speed; therefore, as the photon rises, it will not undergo a change in speed, but it will lose energy. Now, energy is linked to frequency, which for visible light indicates its colour. So, as the photon rises, it loses energy, and its frequency shifts towards lower frequencies, i.e. towards red. If the photon started from the roof of the container and fell downwards, it would gain energy, and its frequency would shift towards blue. As shown by Einstein, the ratio between the frequency measured for a photon that has escaped to infinity and a nearby one is proportional to the mass. This effect is also linked to another phenomenon known as *gravitational time dilation*, i.e. time flows differently in the presence of masses in gravitational fields. The reason is that *gravitational redshift* shows how the frequency changes in a gravitational field, but frequency is one way of measuring time. Frequency changes in gravitational fields are due to the fact that they produce changes over time. So, the origin of gravitational redshift is linked to the way time flows in gravitational fields. To explain all this,

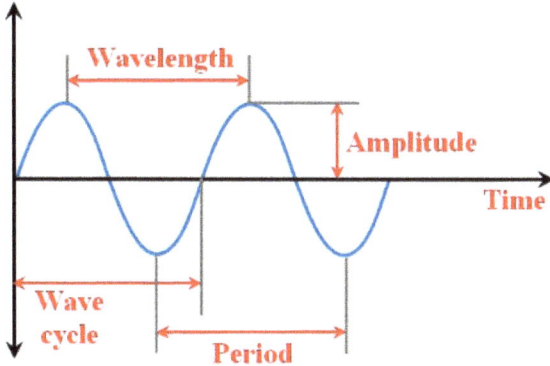

Figure 4.10.   A wave and its parameters.

let's consider the vibration of a wave of light, as in Figure 4.10. Let's consider a clock that ticks every time there is a vibration of the wave. The time, which is the inverse of the frequency, between two ticks is equal to the oscillation period of the wave, $\Delta t = 1/v$, i.e. the time between two maximums, as in Figure 4.10.

Now, as we said, near the Earth, or a mass, if the photon moves away, it will undergo a redshift and, therefore, a decrease in frequency. Since time is inversely proportional to frequency ($\Delta t = 1/v$), it will pass more slowly as the gravitational field or its mass increases. This is the effect called *gravitational time dilation*. We can therefore say that gravitational redshift is produced by the fact that time flows more slowly near massive objects. This effect can be described in an even more intuitive way again using the equivalence principle through the following mental experiment. Consider two observers on a spacecraft. Observer S (Figure 4.11(a)) is located in the tail of the spacecraft and has a clock that sends pulses at regular times. At the front, we have another observer, S', who is equipped with another clock that receives pulses from S and records the arrival time. Suppose that the spacecraft moves forward with a constant acceleration (Figure 4.11(b)).

Its front part moves more and more rapidly forward, moving away from the impulses sent by the observer S. The light travels at a constant speed, but not the spacecraft. The impulses sent from S will take longer and longer to reach S'. If pulses are sent every second from S, the observer S' will record them, say, every two seconds. For S', the clock used by S

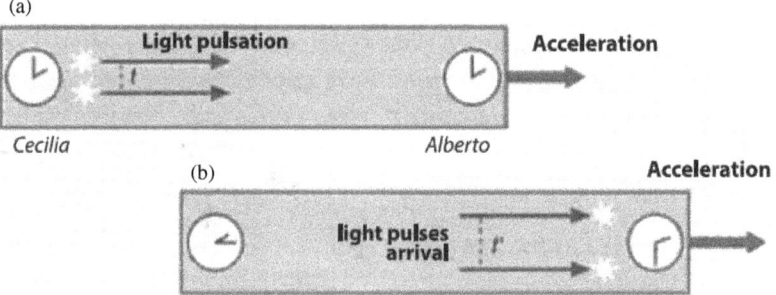

Figure 4.11. The light pulses arrive at different times, depending on the point of the spacecraft from which they were sent.

runs slower. If the impulses were sent from S' to S in the opposite direction to that of motion, the opposite would happen. The rear of the spacecraft moves faster and faster towards the light pulses, and observer S would receive the pulses more and more quickly. For S, the clock used by S' runs faster. By the principle of equivalence, the acceleration is equal to a gravitational field, which exerts its attraction towards the rear part of the spacecraft. So, at the back of the spacecraft, where the gravitational field is strongest, time will flow more slowly, and if we move away from this point to the front, where the field is weakest, time will flow more quickly. Another prediction of general relativity is the emission of gravitational waves (GWs) from moving masses, as happens to electric charges from which electromagnetic waves originate. GWs are perturbations of space-time that move at the speed of light. They were predicted in 1916 by Albert Einstein. However, in 1936, Einstein arrived at results that contradicted those made in 1916 and was convinced that GWs did not exist. There was an error in his calculations, which was corrected after discussions with colleagues. GWs, as mentioned, are generated by the motions of masses and are real ripples in space-time, like the surface of a pond when you throw a stone into it. As we will see, this other prediction of general relativity was also verified.

Are We Sure This is Really the Case?

We ask ourselves again, as in the case of special relativity, the question of whether the predictions of relativity correspond to what is observed in

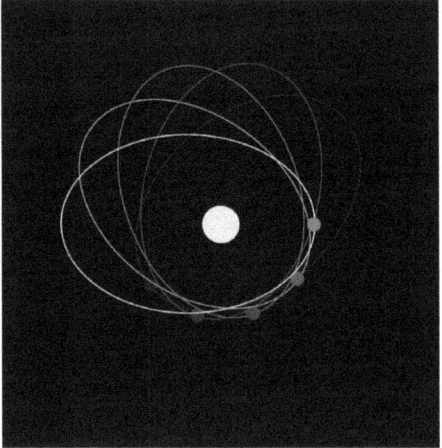

Figure 4.12. Precession of Mercury's perihelion.

Source: Wikipedia, Rainer Zenz.

experiments. The first empirical confirmation of the theory of general relativity was obtained by Einstein himself by explaining an anomaly in the orbit of Mercury. Newtonian mechanics predicts that the orbits of the planets are ellipses with perihelia stationary in space. When a planet is subject to gravitational disturbances, or anomalies in shape, the perihelion[1] can be subject to a precession, i.e. an advance. Mercury, among all the planets in the solar system, is the one that presents the most accentuated perihelion precession, being the closest to the Sun.

This was noticed by Urban Le Verrier. Mercury's perihelion precession is 574 arcseconds per century (Figure 4.12). After subtracting 531 arcseconds due to disturbances generated by other planets, 43 arcseconds remained. Le Verrier considered them evidence of a planet orbiting between the Sun and Mercury, and it was called Vulcan. The absurd thing is that in 1978, James Watson and Louis Swift announced the observation of Vulcan, which however, being non-existent, was not confirmed by other observations. Einstein showed that the 43″ of arc were produced by relativistic effects. General relativity predicts that Newton's force, in addition to having the classic term dependent on the inverse of the square of

---

[1]Perihelion is the maximum distance of a planet from the Sun, while aphelion is the minimum distance.

the distance, contains an additional term dependent on the inverse of the cube of the distance. The gravitational field produces more gravity at the perihelion than at the aphelion. Consequently, the orbit does not close, and there is a rotation of the major axis of the orbit which explains the precession of 43″ of arc.

We have seen that gravity deforms space and that light is deflected. The first attempt to verify the prediction regarding the deflection of light was made by Erwin Freundlich, who organised an expedition for the eclipse of August 21, 2014, in Crimea. The expedition was not successful because the members of the expedition were captured by the tsarist troops during the war. The event worked in Einstein's favour. In fact, at that time, his theory predicted a deflection of 0.87 arcseconds,[2] equal to that calculated by von Soldner using the corpuscular theory of light and Newton's theory of gravitation. Einstein made a second estimate, made public at the famous conference of November 25, 1915. That estimate, 1.75 arcseconds, was about double what was predicted the previous year. Obviously, the mishap that happened to Freundlich was of great benefit to Einstein since if the mission had been successful, the results would have contradicted his theory. The second attempt to verify "the most beautiful of theories" was that of the director of the Cambridge Observatory, Arthur Eddington, who organised a double expedition to the island of Principe in Africa and to Sobral in Brazil. In addition to the desire to verify the theory, the expedition had a more "practical" objective, that of avoiding Eddington being called up for military service. As a pacifist Quaker and conscientious objector that he was, he would certainly have refused to enlist, and this would have also been an embarrassing situation for Trinity College, for the Cambridge Observatory, and consequently for English science. Astronomer Royal Franck Dyson pointed out to colleagues that, on May 29, 1919, there would be an eclipse that would cross the Atlantic and that it was particularly interesting for verifying Einstein's theory since the Sun and the Moon would be found in the constellation of Taurus, in whose centre the Hyades cluster is located. It was an almost unique opportunity, and it killed two birds with one stone: verifying Einstein's theory and solving the "Eddington" problem. Under Dyson's urging, everything was organised, and the two expeditions set off aboard the *Anselm*, His Majesty's ship. Despite the meteorological problems, the two missions

---

[2]More precisely 0.83″ because there was a calculation error involved.

took photographs used by Eddington to determine the value of the deflection, after having discarded data that deviated from his expectations. The angular deviation predicted by general relativity is tiny, corresponding to the angle subtended by a finger at a distance of 1 km.

The result was declared to be in accordance with the predictions of Einstein's theory, which immediately became famous throughout the world. We have seen that the light deflection effect produces the distortion of distant galaxies, the formation of multiple images, or rings, due to the presence of an object between the observer and the source and is known by the term "lens effect". The first studies on the lens effect were carried out by Orest Chvolson in 1924 and by Einstein in 1936. In reality, Einstein had already derived the effect in 1912, three years before the publication of the theory of general relativity, as a consequence of the deflection of light by gravitational fields, but the result had not been published.

Einstein had written the results of the calculation in a notebook that he had brought with him when he went to Berlin, where he met the astronomer Freundlich. He had discussed some possibilities with him to test his ideas. In 1936, he was visited in Princeton by a Czech dishwasher, Rudi Mandl, who discussed the gravitational lens effect with Einstein. According to Mandl, starlight focused by a gravitational lens could have influenced human evolution, producing important genetic mutations. Under Mandl's pressing request to publish an article on the lens effect, Einstein, who had initially refused because, in his opinion, the effect was negligible, published it in an article in *Science*. Curiously, in the article, he explicitly wrote that he had published the article to satisfy Mandl's wish. Astronomer Fritz Zwicky was the first to indicate that galaxy clusters could behave like gravitational lenses. This prediction was published in a 1937 article in which Zwicky applied the virial theorem to the Coma cluster. More than 40 years had to pass before his intuition and the predictions of general relativity were verified in 1979, when an object duplicated by the said effect was observed, the twin quasar SBS 0957+561. Other confirmations of the gravitational lensing effect come from the observations, made by Lynds and Petrosian, of gravitational arcs produced by the distortion of images of galaxies behind a cluster.[3]

---

[3] Article written in 1970 and published in 1986.

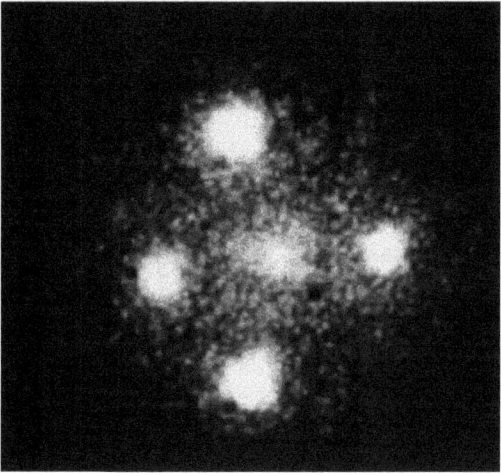

Figure 4.13.   Einstein cross produced by the galaxy ZW 2237 +030 and the Quasar G2237 +0305.

Source: ESA/Hubble and NASA.

In the case of source, lens, and observer alignment, the following can be observed:

- four copies of the source, called the Einstein cross effect: an example of the effect is produced by the galaxy ZW 2237 +030 and the quasar G2237 +0305 located directly behind it (Figure 4.13);
- a ring, the Einstein ring, the first of which was discovered in 1998: B1938+666 (Figure 4.14).

The arcs and rings produced by the lens effect are simply cosmic mirages, or optical illusions. They are similar to mirages observed in deserts, in that case produced by temperature differences between the ground and the air, which deflect the light. In astronomy, the lens effect and its various manifestations have great importance in determining the mass distribution of astronomical objects. From the distortion of distant galaxies, it is possible to determine the total mass of the lens.

As regards the *gravitational redshift*, this prediction was verified by Robert Pound and Glen Rebka in 1959. The two used the tower of the Jefferson Laboratory at Harvard University to install an emitter, consisting of an unstable isotope of iron that emitted gamma rays, positioned at

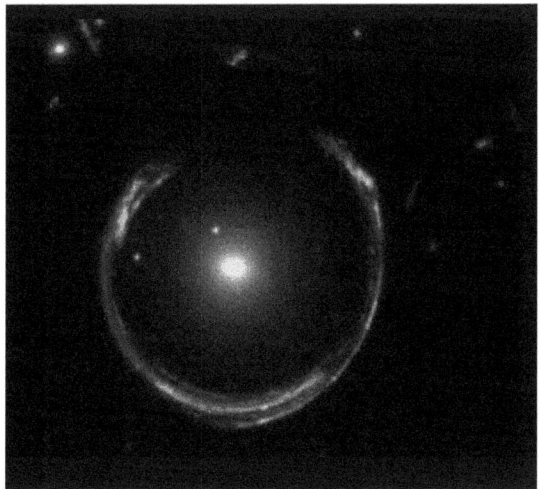

Figure 4.14.   Einstein ring.
Source: ESA/Hubble NASA.

the top of the tower, while a receiver was positioned at the base of the tower. The experiment showed perfect agreement with the predictions of general relativity. At the base of the tower, time passed 210 trillionths of a second slower than at the top of the tower. Subsequent, more precise experiments showed agreement between experiment and theory within 0.007%. As regards the time dilation produced by gravitational fields, it was verified in an experiment carried out by Joseph Hafele and Richard Keating in 1971. The experiment is actually a verification of both special and general relativity. The two physicists took atomic clocks onto airplanes. The clocks circled the Earth twice in opposite directions. Upon returning to Earth, the times measured were compared with each other and with the time marked by a watch left on the ground. The clock that travelled east was 60 billionths of a second behind the one on Earth,[4] while the clock that travelled west was 270 billionths of a second ahead, in accordance with the predictions of relativity, taking into account the time dilation due to altitude and the dilation due to the motion of the clocks predicted by special relativity. In 1976, Luigi Briatore of the University of

---

[4]In the flight towards the east, the speed of the plane adds to that of the Earth's rotation, and the kinematic effect prevails over the gravitational one, which is of the opposite sign.

Turin and Sigfrido Leschiutta of the Galileo Ferraris Institute verified gravitational dilation by comparing the times measured by two clocks, one located in Turin at a height of 250 m and another on the Plateau Rosa at a height of 3,500 m. The two clocks showed a difference of 34 billionths of a second per day, in agreement with general relativity. As already mentioned, the global positioning system (GPS) is a further confirmation of both special and general relativity. We have already talked about the corrections related to special relativity; as regards those relating to general relativity, it must be remembered that a clock on Earth feels a much greater gravitational field than an atomic clock on a satellite, which will be accelerated by approximately 45 billionths of a second per day. Multiplying by the speed of light, we find that after one day, we would have an error of about 14 km. Overall, the atomic clock on the satellite is slowed down by about 7 billionths of a second per day due to special relativity and sped up by about 45 billionths of a second per day due to general relativity. Ultimately, it is accelerated by about 38 billionths of a second per day. Without taking these effects into account, the GPS would produce an error of approximately 11 km. Other verifications are based on lunar laser telemetry. The Apollo missions left mirrors on the moon. Astronomers at the Apache Point Observatory in New Mexico directed laser beams at the reflectors. In this way, it was possible to measure the lunar motion with a precision of 1–2 cm and verify whether general relativity correctly predicts the motion. The motion of the Moon occurs as predicted by general relativity to an accuracy of one part in ten trillion. Another fundamental confirmation of general relativity is the discovery of GWs on September 14, 2015, by LIGO (Figure 4.15).

The latter consists of two detectors, one in Louisiana and the other in Washington State. They are two interferometers with two L-shaped arms 4 km long. The passage of the 2015 wave caused a tiny variation in the length of the arms (a factor of the order of one ten thousandth of the radius of a proton), which was measured anyway. Previously, in 1974, Russell Hulse and Joseph Taylor indirectly confirmed the existence of GWs by studying the decay of the orbit of the binary pulsar PSR 1913+16 due to the emission of GWs. The distance between the two pulsars decreases by 7 mm per day. The two stars are expected to collide in 85 million years, producing an explosion with the emission of gamma rays. From all the confirmations, general relativity appears to be one of the most precise and important theories in physics. This has both positive and negative aspects. It's good to have a theory that predicts gravity so precisely, but open

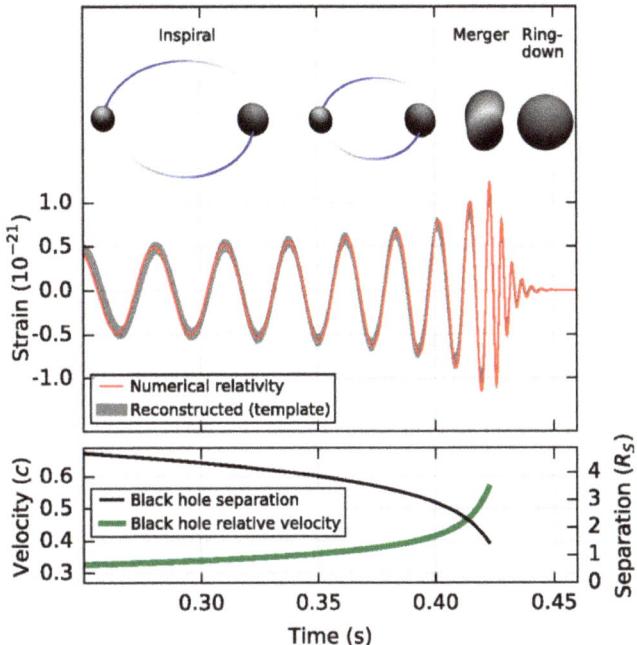

Figure 4.15. The event called GW150914, being a GW received on September 14, 2015, due to the merger of a 36 $M_\odot$ black hole and a 29 $M_\odot$ black hole. The top panel shows how the black holes approach each other and fuse into one black hole. The oscillating line denotes the gravitational-wave strain amplitude. Bottom panel: black hole relative velocity and separation.

Source: Ligo Collaboration (https://doi.org/10.1103/PhysRevLett.116.061102).

questions remain. To describe the Universe in its very early phases or black holes, a new theory is needed that combines general relativity and quantum mechanics. Despite decades of effort, no such theory still exists. Furthermore, the dynamics of galaxies, galaxy clusters, and beyond are not correctly described by general relativity, and it is not clear whether this is due to the need for another theory of gravity or the fact that, in the Universe, a large part of the mass is not observable (i.e. dark matter). Relativity gives us a paradoxical, not very intuitive, vision of reality, but in this it is in good company: quantum mechanics shows us an even stranger world. We can only adapt to the idea that the world is like this.

Chapter 5

What is a Black Hole?

*Black holes ain't as
black as they are painted.*

— Stephen Hawking

Black holes, among astronomical objects, are among those that arouse the most curiosity and interest. They have often been used by science fiction writers in their books, but unlike many other ideas that have little to do with reality, they are profoundly real objects to which science has dedicated many studies. Black holes are the result of the force of gravity, which, with its tendency to attract different masses, manages to compress large objects, such as stars, into narrow spaces. The first to be interested in it was the English philosopher and clergyman, John Michell. After studying at Cambridge, he taught both literary and scientific subjects. At age 43, he became rector of the town of Thornill. Since work did not occupy him much and he was in a good economic position, Michell left university and continued to pursue his interest in science. He became the first modern scholar of earthquakes, studied light and magnetism, designed the experimental apparatus that was used by Cavendish to measure the gravitational constant, and was the first to use statistical methods for astronomy and what are now known as black holes. In November 1783, he sent a letter to the Cavendish laboratory, later published by the Royal Society, which proved prescient. Michell's basic idea was to measure the mass of a star. Like Newton, Michell thought that light was made up of corpuscles and that, therefore, the speed of light exiting the stellar surface must be slowed down by the force of gravity. In order to calculate the mass of a star, according to him, it was enough to measure the speed of light of the stars. Michell used James Bradley's measurement of the speed of light, which was a value of 301,000 km/s – very close to the exact value – and answered the question of what happened to a star so massive that it has an escape velocity equal to that of light. *Escape velocity* is the minimum velocity that must be imparted to an object in order for it to escape another object's gravity. For example, in the case of the Earth, this speed is 11.2 km/s, and in the case of the Sun, 617.3 km/s. So, for light, at approximately 300,000 km/s, it is easy to escape from terrestrial or solar gravity. If an object were so massive that it had an escape velocity equal to that of light, it would no longer be visible. In his letter, Michell concluded:

> *If the half-diameter of a sphere of the same density as the Sun in the proportion of 500-1, and supposing that light is attracted by the same force in proportion to its [mass] as other bodies, all the light emitted by such a body would be was made to return towards it, by its own gravity.*

The same idea was proposed a few years later by Pierre-Simon de Laplace. This result was considered by the two as a theoretical artefact, and the scientific community itself did not believe that light could be captured. For the sake of precision, it should be noted that Michell's idea that gravity can cause light to slow down is in contradiction with Einstein's ideas of special relativity, but Michell was right in supposing that any body with an escape velocity greater than the velocity of light must be invisible. The idea fell into oblivion and was revived a couple of centuries later when Einstein, with his theory of general relativity, showed that gravity was nothing other than the curvature of space-time and that the trajectory of light would also be modified in the presence of this curvature. A few months after the publication of general relativity, Karl Schwarzschild applied the theory's equations to a non-rotating, spherical object and found that celestial objects with gravity so intense that they could trap light could exist.

The dimensions of such an object depend on the typical radius of the *Schwarzschild solution*, called the *Schwarzschild radius*, which is proportional to the mass of the object. To give some examples, in the case of the Earth, this radius would be approximately 0.9 cm, and in the case of the Sun, approximately 3 km. The Schwarzschild metric concerns a massive, non-rotating object. This solution to the equations of general relativity was generalised in 1963 by Roy Kerr to the case of a non-charged object, and in 1965, an extension for a charged body was obtained, the so-called *Kerr–Newman solution*. From the previous discussions, it is clear that the fundamental difference between a black hole and a star is all a question of density. If, for example, we were able to compress the mass of the Sun, which is 2×10^{30} kg, and its radius (696,000 km) into 3 km, a black hole would be generated. The question that arose was whether this was feasible. The solution obtained by Schwarzschild was considered for decades to be a mere mathematical curiosity. The question arose of how a celestial object could be compressed to such tiny dimensions. In 1939, Einstein himself wrote an article in which he argued that it was impossible for a star to collapse under the action of gravity because the matter would resist being compressed beyond a certain limit. To truly understand what a black hole is, it is necessary to have clear ideas about what gravity, a force we all know, is capable of doing. It keeps us glued to the Earth, and if we throw a stone into the air, it falls back, similar to Newton's famous apple, which gave the famous scientist the idea to formulate his theory on gravitational force. The most interesting thing is, as Newton himself intuited, that the same force, in addition to regulating the fall of bodies, regulates the motions of the planets in the sky. Although it is the weakest of the known forces of nature, it has two characteristics that make it successful over great distances and times. First, it has an infinite range of action, and second, it is always attractive, unlike, for example, the electromagnetic force, which is both attractive and repulsive, and this limits its action over large distances.

Three Ways to Die

Today, we know that Einstein's idea that it is impossible for a star to collapse under the action of gravity was wrong. The life path of a star depends fundamentally on its mass. A star is formed by the gravitational collapse of dense, cold clouds of molecular gas. As the collapse increases, the cloud shrinks, and its central temperature increases until it reaches values sufficient to trigger nuclear fusion, specifically the fusion of four hydrogen nuclei to form a helium atom. Energy is released in the process. The star is in equilibrium under the action of gravity, which tends to make it collapse, and that of gaseous and radiation pressure, which tends to make it expand. This equilibrium continues as long as fusion reactions continue. In the case of stars with a mass lower than 8 solar masses, when a good part of the hydrogen in the core is consumed, the radiation pressure that balances gravity decreases, and the central part of the star contracts, triggering the hydrogen in a shell around the centre.

Due to higher temperatures, the rate of nuclear reactions is greater, and this causes the star to increase in brightness by a factor of between 100 and 1,000. The increase in the density of the core and its temperature translates into an expansion of the surface layers of the star. Because the energy produced is released over a larger surface area and because some of it is dissipated in the expansion, this results in a lower surface temperature of the star. The star is now a *red giant*. The life of the star is prolonged by the ignition of helium in the core.

When the helium also runs out in the core, there will be a new expansion of the outermost layers and a contraction of the innermost ones, allowing the helium to fuse into a shell around the centre. However, stars with a mass lower than 8 solar masses do not have sufficient mass to reach the temperature and pressure necessary to fuse carbon, and the star, no longer supported by radiation pressure, will collapse under its weight and expel most of its mass, forming a *planetary nebula*. Only the core will remain to form a star known as a *white dwarf*, with a mass similar to that of the Sun, a size a thousand times smaller, and a density a million times higher than that of the Sun. The collapse is formed at this level by the fact that, according to quantum mechanics, electrons cannot come together indefinitely. To understand the equilibrium of white dwarfs, we need to remember two of its principles: the *Pauli exclusion principle* and *Heisenberg's uncertainty principle*. The exclusion principle applies to particles that make up ordinary matter: electrons, protons, and neutrons, and states that these particles cannot simultaneously occupy the same quantum state. What does it mean? If we consider a hydrogen atom, it is made up of a proton at the centre and an electron that moves around the proton in regions called *orbitals*. The state of an electron is defined by four numbers, called quantum numbers: the *principal quantum number* (related to energy), the *angular momentum quantum number*, the *magnetic quantum number*, and the *z component of the spin*. Spin, although a misnomer, is often referred to as a rotation around the axis and can be clockwise or anti-clockwise. The first orbital of the hydrogen atom has a spherical shape. Suppose that an electron moves in this region. It has three values for the first three quantum numbers and one for the spin. We mentioned that there are two possible values of the spin, clockwise or anticlockwise; therefore, on this orbital, we can place another electron with a spin opposite to that of the other electron. However, we cannot add a third one because it would have the same quantum number values as the others. The third must go to another orbital, further away from the proton. The Pauli principle explains the stability of matter. The electrons cannot be included in any small regions but must be at a certain distance from each other. Two electrons or two neutrons cannot be located close together due to the Pauli exclusion principle. The second principle, *Heisenberg's uncertainty principle*, states that it is not possible to simultaneously know with infinite precision the position and velocity of a particle: the better we know the position of a particle, the worse we know its velocity, and vice versa. How can these principles of the microscopic world influence the state of a star? As we have seen, when nuclear reactions stop, the star tends to collapse and gravity pushes the atoms close to each other, and the same thing happens to the electrons. Even if the electrons are not in their orbitals, due to the exclusion principle, they cannot get closer at will. They are then compressed to a minimum distance. Therefore, in the lower-energy state, the electrons fill all the levels up to a maximum, called the *Fermi energy*, below which there are N states. The volume within which the electron can be found is a sphere with a radius of 2.4×10^{-10} cm. We can calculate this volume in a simple way. We know that a white dwarf has a radius between 0.008 and 0.02 that of the Sun, i.e. between 5,600 and 14,000 km. Let's take an intermediate value, 10,000 km, which corresponds to a volume of $4\pi/3 \times$ radius^3, i.e. 4×10^{21} m^3. The mass of a white dwarf is between 0.5 and 0.7 solar masses, with a peak around 0.6. Let's assume, for simplicity, that it is 1 solar mass. In one solar mass, there are approximately 10^{57} atoms. The volume that belongs to each atom is given by the ratio between the volume and the number of atoms: $4 \times 10^{21}/10^{57}$ which is 4×10^{-36} m. This corresponds to a cube with a side of approximately 10^{-10} cm, which is the typical size of an atom. In the white dwarf, atoms and electrons are tightly packed. Now, let's remember the uncertainty principle. When we force electrons into a small space, this means that we know their position well, and therefore the speed must be indeterminate and high. That is, in other words, the electrons are close together, but they all move at very high speeds. This produces a pressure, called *degeneracy pressure*, which counteracts the force of gravity. This pressure has this name because the state of matter in a white dwarf is called *degenerate matter*, which has very high density characteristics. Although white dwarfs have masses in the range of 0.5–0.7 solar masses, there is an upper limit for their mass, corresponding to 1.4 solar masses, called the

Chandrasekhar limit. This limit was found by a young Subrahmanyan Chandrasekhar in 1930 on his journey on a ship that took him from India to England. In 1935, Chandrasekhar presented his result, but Arthur Eddington, at that time considered the founder of modern astrophysics, took the floor and attacked the results by saying, among other things, that "I believe that there must be a law of nature that prevents a star from behaving in this absurd way". After a short time, Chandrasekhar moved to America, a less hostile environment, and in 1983, he was awarded the Nobel Prize in physics for his studies on the structure and evolution of stars. History showed that Chandrasekhar was right and Eddington was wrong.

We have seen some of the characteristics of a white dwarf. They are stars that contain a mass close to that of the Sun in a volume one million times smaller and consequently have a very high density of the order of one ton per cubic centimetre. They have surface temperatures of 100,000 degrees, and for this reason, they appear white even if their brightness is low. The interior of the white dwarf is made up of atomic nuclei of ionised carbon, oxygen, neon, magnesium, and helium. These nuclei are surrounded by a sea of free electrons, which, thanks to the degeneracy pressure, support the structure of the star. The star cools slowly, over billions of years, until the material crystallises. A characteristic of white dwarfs that distinguishes them from normal stars is that the radius of a white dwarf is inversely proportional to the cube root of its mass. So, the more massive the white dwarf, the smaller its size, unlike normal stars. Today, many white dwarfs are known. The first was discovered by Friedrich Wilhelm Bessel, a great German mathematician and astronomer. Sirius is the brightest star in the sky. Its motion does not follow a straight line but twists in a serpentine motion. This betrays, in accordance with Newton's laws, the presence of a companion not easily visible, at least with the technology of Bessel's time. The companion, called Sirius B, was discovered 20 years later. It was expected that the star, being not very bright, would be red, but instead, in 1915, it was seen that it was white, with a mass similar to that of the Sun. If the mass of the star is greater than 8 solar masses, the temperature and pressure at its centre are so high that they melt elements heavier than carbon. For carbon fusion to begin, temperatures above 720 million degrees and stars with a mass greater than 8 solar masses are needed. Carbon fusion reactions produce several products, mainly sodium, magnesium, oxygen, and neon. Similar to the case of helium, depending on the mass, the fusion of carbon can occur with or without a flash. This effect, the expansion of the star and the increase in brightness, causes the radiation pressure to produce stellar winds, and the star gradually loses mass. Carbon fusion produces an increasing concentration of neon and oxygen in the core. Shells form around the nucleus, in which, going outwards, the fusion of carbon, helium, and hydrogen occurs. The contraction following the cessation of carbon fusion increases the temperature up to a value of 1.4 billion degrees. At this temperature, in stars with a mass greater than 10 solar masses, neon is triggered, and *photodegradation* processes of the neon itself also begin. The neon nuclei, interacting with very high-energy photons, separate into helium and oxygen. These nuclei are subject to a series of reactions: oxygen and helium give rise to neon; neon and helium produce magnesium; and magnesium and helium produce silicon. The products of the reaction accumulate at the centre of the star, and the neon begins to burn in an outer shell. When the temperature reaches 1.8 billion degrees, oxygen fusion reactions occur, which produce elements important for life, such as sulphur and phosphorus. Photodisintegration processes of magnesium begin, and at 3.4 billion degrees, those of silicon begin, leading to the formation of more stable nuclei, such as iron and its isotopes, including ^{56}Fe. The star has an onion-like structure (Figure 5.1).

Burning silicon produces the most stable element in nature, iron-56. If you want to fuse iron to form a heavier element, you need to supply more energy than is released by fusion. When it gets to this point, the

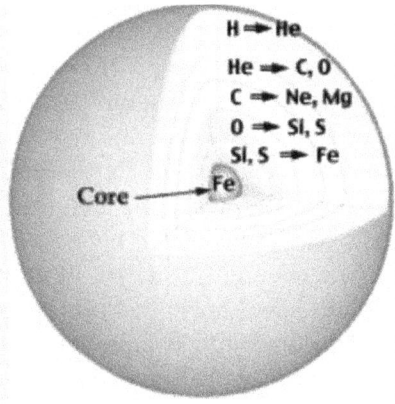

Figure 5.1.   Structure of a massive star.

nuclear energy source at the centre of the star is consumed. Gravity is not hindered by anything and can cause the star to collapse. The contraction produces a huge increase in temperature that reaches 5 billion degrees in just a few minutes. The photons are so energetic that they can photodisintegrate iron, producing neutrons and helium. Since iron fusion reactions require energy instead of releasing it, this produces further cooling of the core, which accelerates contraction. The iron core that forms inside massive stars has a mass greater than that of the Chandrasekhar limit and therefore collapses on itself due to gravity. The energetic photons present split atomic nuclei into neutrons and protons. Over the course of evolution, we went from light elements to increasingly heavier elements, up to iron. Now, we retrace the reverse path: from iron, we return to hydrogen. The protons, due to the high pressures, capture electrons, forming neutrons with the emission of neutrinos. These processes (the breaking of nuclei, the formation of neutrons, and the emission of neutrinos) consume energy and therefore favour the collapse of the nucleus. It collapses under its weight and is slowed down until it is stopped by the degeneracy pressure of the neutrons. The nucleus, made up of neutrons, reaches very high densities, 10^{17} kg/m^3, forming a neutron star of approximately 10 km.

In this case, the star will produce a huge explosion with an intensity that can surpass that of an entire galaxy. This state is referred to as a *supernova*. What happens next is similar to that observed in stars with a mass lower than 8 solar masses: the electrons stop the collapse. This time, the collapse is stopped by the *degeneracy pressure of the neutrons*. If we wanted the Earth to have the same density as a neutron star, it would have to contract to 100 m. The gravitational energy generated by the collapse is very high. The nucleus shrinks from Earth-like dimensions to 10 km in just one second. In this process, an energy of 10^{46} J is released. In 2021, the energy consumed by the European Union was 39,351 PJ (where PJ is petajoule, i.e. 10^{15} J), or 3.9×10^{19} J. In other words, the energy released by the collapse could provide energy to the European Union for 10^{27} years, a time enormously longer than the age of the Universe. The part of the star above the iron core falls onto the core and bounces off it. It was thought that the explosion the star undergoes was produced by this bounce, but supercomputer calculations have shown that this is not the case. A shock wave is formed that moves outwards, compressing the material and triggering nuclear reactions. However, the wave is slowed down by the falling material. The explosion gives rise to a supernova. Just 1% of the energy from the collapse of the iron core is enough to generate the expansion speed of the material in the supernova that forms due to the collapse of the core. It is not yet known how this energy is transformed into the kinetic energy of the material, though it has been suggested that neutrinos play a fundamental role. In fact, at the high densities present in the star, neutrinos interact with matter, transferring their energy to the pressure wave that moves outwards. The pressure wave passes through the entire material in a few seconds, compressing all the layers. In the innermost areas, compression heats the material to sufficiently high temperatures to trigger nuclear reactions and explosive combustions, first in silicon, then in oxygen, then in neon, and finally in carbon. The released nuclear energy accelerates the various layers until they are expelled. In the most central area, the energy is not sufficient to produce this, and the innermost layers fall back onto the neutron star. In the case of a star of 25 solar masses, the neutron star would have a mass of 2 solar masses. The combustion phases, starting with the ignition of carbon, are dominated by neutrinos, which significantly shorten the stellar evolution times. From the carbon fusion phase onwards, enough energy is present to generate electron–positron pairs. These pairs can annihilate, producing energy in the form of photons and neutrinos. The energy loss in the form of neutrinos is hundreds of millions of times greater than that lost in the form of photons. To compensate for this loss of energy, the star produces more energy through nuclear reactions. The star's internal combustion cycles become faster and faster. For example, in a star of 25 solar masses, the central combustion of hydrogen takes six million years, and half a million years is taken for the combustion of helium. The combustion of carbon takes less than 200 years, that of neon less than a year, oxygen is burned in four months, and silicon within a day. This acceleration in burning times is due to the loss of energy in the form of neutrino emission. There is a third possibility for the death of a star: if the mass is higher than the Chandrasekhar limit and the *Tolman–Oppenheimer–Volkoff limit*, i.e. between 2 and 3 solar masses, and when neither the pressure of the electrons nor that of the neutrons can stop the collapse. This situation was studied by Robert Oppenheimer, George Volkoff, and Hartland Snyder. In this case, in nature, there is no force that can stop the collapse, and hence the collapse will not stop and will continue up to a point of infinite density. This point is called a *singularity*. At a singularity, the curvature of space-time becomes infinite, and the theory of general relativity no longer works.

In the following years, research stopped because American, Russian, and other scientists participated in the Manhattan Project for the development of the atomic bomb. In 1958, John Wheeler became interested in the problem of the implosion of a star and championed the notion of a black hole. Wheeler's team, in 1959, suggested that the internal structure of a black hole is governed by quantum gravity, a still incompletely developed theory that combines gravitation with quantum mechanics. Wheeler proposed a famous theorem called the *no-hair theorem*, also known as the *essentiality theorem*.

According to this theorem, it is impossible to say what is inside a black hole. All the peculiarities of the collapse of a star to form a black hole are lost in its formation. Black holes are all similar, regardless of how they formed. The most general black hole is described by only three quantities: its mass, its charge, and its angular momentum, linked to rotation. The Russians also contributed to the development of knowledge of black holes. In 1961, Yakov Zeldovich formed his own astrophysics and general relativity group. The group made major contributions to the thermodynamics of black holes. In 1964, the golden period of research on black holes began, during which Hawking, Penrose, Rees, and Carter were involved. In 1973, Hawking and Kip Thorne met Zeldovich and Alexej Starobinskij in Moscow, who explained how rotating black holes emit particles and also how, due to the uncertainty principle, black holes had to emit and absorb particles. These studies also reverberated in science fiction. In 1967, in an episode of Star Trek, the ship *Enterprise* was in danger of falling into a black star. In the same year, Wheeler coined the term *black hole*, which replaced the term *frozen stars*, and it immediately became popular.

Where are Black Holes Found?

If black holes were physical entities, as demonstrated by many studies, the problem remained in understanding how to observe something that emitted neither light nor other forms of radiation. The Russian Yakov Zeldovich had the right idea. Stars lose mass during their lifetimes. If a star has a companion that is a black hole, it could gravitationally attract gas from the star, causing the gas to orbit around it, forming an *accretion disc*. In this way, the gas would reach temperatures of millions of degrees and consequently emit radiation in the X-ray part of the electromagnetic spectrum. So, it was necessary to look for binary systems with optical and X-ray telescopes. This way, the optical telescope would reveal the star, while the X-ray telescope would reveal the black hole. These objects are

Figure 5.2.   Artist's image of a black hole with accretion disc and jet.

called *X-ray binary systems*. As mentioned, they are made up of a normal star that loses mass in favour of the black hole around which an accretion disc forms. Through this, the matter falls towards the central object after losing its angular momentum (Figure 5.2). The material rotating around the black hole loses angular momentum and, due to friction, heats the disc and transfers a large part of the gravitational potential energy released towards the external regions. Using Zeldovich's idea required waiting for X-ray telescopes to be put into orbit in the early 1970s. The first source observed was located 8,000 light years from Earth and had a mass greater than seven times that of the Sun. Analysis of the object led to the conclusion that the X-ray emission came from a cloud of superheated gas falling in a spiral towards a black hole, renamed Cygnus X-1. There have been many systems like Cygnus X-1 observed since then, such as the system consisting of the star HDE 226868 of 30 solar masses and a black hole with a mass between 5 and 10 solar masses. The gas falls from the star in a spiral orbit onto the black hole, forming an accretion disc. Magnetic fields compress the gas and expel it in jets of plasma[1] in a direction perpendicular to the accretion disc, which emits radio waves.

---

[1] Plasma is a state of matter different from solid, liquid, and gaseous states. It behaves like a gas and is often described as an ionised gas at very high temperatures.

There are black holes of different sizes. Stellar black holes have masses on the order of a few solar masses and are generated by the collapse of a massive star into a supernova. There are also black holes of intermediate mass, between 100 and 1,000 solar masses. It is not known exactly how they form, but they could originate from the merger of stellar black holes. Then, there are *supermassive black holes*, with masses between millions and billions of solar masses. They are located at the centre of galaxies. In 2015, with the first observation of gravitational waves, black holes with a mass of around 30 solar masses were observed, but it is not known how they originated. Finally, as we'll discuss later, there are so-called *primordial black holes*.

Galactic Monsters

Inside every galaxy lies a giant black hole. In most cases, as in our galaxy, this galactic monster seems dormant. It is not known how galactic black holes form. According to some theories, they form before the galaxy, whereas in others, the galaxy forms first, and in still others, their evolution occurs in an almost symbiotic way. The link between the galactic black hole and the galaxy is demonstrated by the close relationship between the galactic black hole and the mass of stars in the galaxy. Among all the galaxies, there are some particular ones called *active galaxies*. Their characteristic is that they have a nucleus (called the active galactic nucleus, or AGN) hundreds of times brighter than the rest of the galaxy. In addition to this characteristic, AGNs, unlike other galaxies, have an emission spectrum that extends across the entire energy spectrum, i.e. the energy is not produced by stars as in normal galaxies. They are very compact, and near the AGN, there is gas that has very high-speed motions, up to 10,000 km/s. These properties can be explained by the presence of a *supermassive black hole* inside the AGN, which would emit a large amount of energy associated with gravitational processes. Supermassive black holes, similarly to the case of stellar black holes, give rise to the formation of a disc of gas and dust rotating around them. Near a black hole and up to about 1,000 *Schwarzschild radii*, the structure is similar to that of a toroidal-shaped structure made of ions. Going outwards, between 1,000 and 100,000 *Schwarzschild radii*, it has a thin disc, which then "pulses up" into small clouds.

At distances of the order of 0.5–1 light years from the black hole, there is a region made up of gas and dust with temperatures between

40 and 60 K[2] and which emits in the infrared. At distances less than 0.1 light years, temperatures reach 1500°C and dust evaporates. From this area towards the interior, the emission is in the ultraviolet and visible bands. If you approach the black hole, at distances of a few light hours, the temperatures are in the order of millions of degrees, and there is emission similar to the case of stellar black holes. AGNs are characterised by jets that emit radiation in the radio band. The brightness of active galaxies can be up to 100,000 times that of our galaxy, with a nucleus 100 times brighter than the galaxy. As a result, the central luminosity dominates that of the galaxy, and active galaxies have a stellar structure. In observations made during the 1960s, AGNs were mistaken for stars in our galaxy and called quasars. It was understood that we were dealing with distant and very bright galaxies. The "condition" of quasars is typical of each galaxy in some phase of their evolution, with a duration much shorter than the typical lifespan of galaxies. The quasar phase is triggered when the supermassive black hole has enough material it can devour. When this material decreases, the quasar transforms into a normal galaxy. Thus, it is expected that the quasar phase may be an intermittent phase in the life of a galaxy.

Falling Into a Black Hole

A question that some people ask is, What would happen if you fell into a black hole? Certainly, falling into a black hole is a negligible eventuality; we don't have black holes close to Earth. These types of questions certainly arise from the curiosity that black holes arouse. Answering this question is possible within the limits of the validity of general relativity. We don't know what's inside a black hole or to what extent general relativity works, but we can certainly describe what happens when we get close to the black hole. From the discussions so far, it should now be clear that a black hole is nothing more than a region of space in which mass has been squeezed to very high densities, and that this mass gives rise to an intense gravitational field. For simplicity, let's consider the simplest possible black hole, the *Schwarzschild black hole*. If we really want to be able to enter a black hole, we will have to choose it carefully. Why? Because the tidal forces, if too great, would stretch us like spaghetti, that is, we would

---

[2]The K, or degree Kelvin, is defined so that zero degrees kelvin corresponds to −273.16 degrees centigrade. The relationship between the two scales is: °C + 273.16 = K.

be, as they say in jargon, *spaghettified*, after which we would be dismembered and reduced to particles. Tidal forces are also present on Earth. For example, if we stand, the force of gravity will be stronger on our feet, which are closer to the centre of the Earth, than on our head. On Earth, the force of gravity is not very strong; therefore, we do not feel this difference in pull. On an object with gravity much greater than that of the Earth, we could perceive the difference in traction experienced by the head and feet, and in or near a black hole, the traction would obviously be very strong. In the case of a Schwarzschild black hole, the dimensions are given by the Schwarzschild radius, which coincides with the event horizon, the radius within which light can no longer escape. The size of the event horizon grows with the mass, M, of the black hole, while the threshold beyond which we would undergo spaghettification grows with the cube root of the mass. This means that for small black holes, such as stellar ones, the spaghettification threshold lies outside the black hole, i.e. its *event horizon*. In a large black hole, such as the one at the centre of a galaxy, the spaghettification threshold lies inside the black hole. To give some examples, in the case of a black hole of 10 solar masses, the event horizon is about 30 km, while the spaghettification threshold is located at 320 km, far outside the event horizon. In the case of a black hole of one million solar masses, the event horizon would be located at 3 million km, while the spaghettification threshold at 14,900 km, much closer to the centre.

So, if we really want to enter a black hole, we need to choose a very large one. Suppose we move parallel to the accretion disc. When we reach a distance equal to three times the event horizon, the accretion disc ends; this is because matter begins to fall quickly towards the black hole. Proceeding up to two and a half times the event horizon, we would arrive at the *shadow area* of the black hole. Due to the distortion of space by gravity, the black hole creates a shadow larger than its size. At a distance equal to one and a half times the event horizon, the light would appear trapped in a circular orbit around the black hole. Space is so curved that it forces light to move in that orbit. Going further inwards, not even light can orbit anymore, and the shadow of the black hole widens more and more, narrowing the image of the universe behind us. Upon crossing the event horizon, we would enter the black hole without noticing anything in particular, but we could no longer go back, and we can only move forward towards the centre of the black hole. If the theory of relativity continues to hold inside the black hole, what would happen to us is that we would inexorably continue our motion towards the centre. Then, we would reach the spaghettification radius and be torn into smaller and smaller pieces until we became elementary particles.

This is the standard scenario. There is some speculation about what might happen beyond the event horizon. One of the various possibilities is that the black hole is connected through a space-time tunnel to a white hole, that is, an object with characteristics opposite to those of a black hole and from which one can only exit. This is what would happen to us, but if someone watched us fall into the black hole from the outside, the show would be very different. To understand what they would see, we need to remember that time is influenced by motion and gravitational fields. To understand this, let's consider two identical twins, one on Earth and one leaving on a spaceship at a speed equal to 90% of the speed of light. If the latter returned to Earth after 50 years, only 20 years would have passed for him, and he would be younger than his twin. The same phenomenon happens near very dense objects, such as a black hole. So, those who look at us from afar would see that we are moving towards the black hole, but the closer we get, the more they would see our motion slowing down. When we reach the event horizon, they would see that we are no longer moving, and furthermore, our image would slowly become redder and redder. This last thing happens because the gravitational field stretches the wavelengths of photons from visible light towards infrared.

Photos of Black Holes

We have detailed theories of black holes: non-rotating, rotating, and charged, and there is indirect evidence for their existence from X-ray binaries. You might wonder if there is direct evidence for their existence. Well, after decades of studies, direct evidence of the existence of black holes has been obtained. Today, there are two images: one published in 2022 of the supermassive black hole at the centre of our galaxy, Sagittarius A*, at a distance of 26,000 light years, having a mass of approximately 4.4 million solar masses and a radius of 13 million km; and a second image, published in 2019, of another supermassive black hole at the centre of a huge elliptical galaxy, M87, located in the Virgo cluster (Figure 5.3).

The 6.6 billion solar mass black hole is located 55 million light years away. These images were obtained using the Event Horizon Telescope (EHT), a group of eight radio telescopes operating on a planetary scale. The telescopes used to achieve this were Alma, Apex, the Iram 30 m telescope, the James Clerk Maxwell telescope, the Alfonso Serrano telescope,

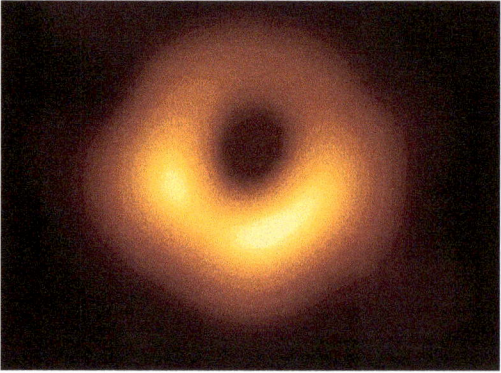

Figure 5.3.   Image of the M87 black hole obtained using data from the Horizon Event Telescope.

Source: Event Horizon Telescope collaboration.

the Submillimeter Array, the Submillimeter Telescope, and the South Pole Telescope. By connecting these eight telescopes located at different parts of the planet, the EHT creates a virtual telescope with dimensions equal to those of the Earth. The EHT and its images are the result of years of collaboration between various groups of over 200 researchers spread across our planet. The data obtained with telescopes and used to obtain the image, for example of the M87 black hole, can amount to millions of gigabytes. These data obtained from the telescopes were recombined by supercomputers at the Max Planck Institute for Radio Astronomy and the MIT Haystack Observatory.

A black hole immersed in a luminous region, such as that produced by the accretion disc, produces, as we said before, a sort of shadow, an effect predicted by general relativity. The shadow is produced by gravitational curvature and the fact that light is retained by the event horizon. Using a variety of imaging and calibration methods, the images showed a ring-shaped region, within which lies the black hole's shadow. The images obtained are extremely important not only because they show that black holes exist, but they can also be studied to verify the predictions of general relativity and obtain information about the behaviour of black holes of different masses. The image of the black hole in our galaxy also allows us to establish the orientation of its rotation. Unlike the black hole of M87 and others, our black hole does not appear to exhibit jets. Furthermore, our black hole does not devour everything that passes by it but is almost dormant, at least for now. However, it has had more active phases in the past. The result that clearly shows the existence of black holes proves Einstein's scepticism about the existence of objects predicted by his own theory wrong.

The Black Hole Paradox: Black Holes Aren't That Black

So far, we have talked about the macroscopic characteristics of black holes, and we have said that they are objects from which not even light can escape if we are inside the event horizon. In reality, in the 1970s, Stephen Hawking showed that this is not exactly the case.

Initially, Hawking discovered a theorem relating to the area of a black hole. He showed that when matter falls into a black hole, the area of the event horizon increases. Furthermore, when two black holes collide and become one, the area of the event horizon is greater than the sum of their areas. The fact that the area could only grow suggested an analogy between the area of the event horizon and another physical quantity that tends to grow: entropy, the physical quantity that indicates the state of disorder of a system. Despite this similarity, it was not clear how the area of the event horizon could be identified with entropy. A Princeton doctorate, Jacob Bekenstein, explained the connection. Initially, Hawking did not agree with Bekenstein's conclusions that a black hole has finite entropy proportional to the area of the events since, in this case, the black hole had to have a finite temperature and emit radiation. Despite initial disagreements between Hawking and Bekenstein, in 1974, Hawking showed that even black holes can emit thermal radiation. The smaller the mass and size of black holes, the greater the radiation. For example, a black hole with a solar mass would have a radiation temperature of 0.0000004 K. If it had the mass of Earth, the temperature would be 0.02 K. In the case of a supermassive black hole, such as those at the centre of galaxies, having masses of a billion suns, the temperature would be 10^{-16} K, while for a small black hole of 10^{18} kg (the size of a mountain), the temperature would be 100,000 K. What causes the emission of radiation from a black hole? In the classical vision of general relativity, we said that it should be dark, and not even light could escape it. We find ourselves faced with a sort of paradox, a situation that often occurs in quantum mechanics. In fact, to arrive at this result, quantum mechanics must be used to redefine the concept of vacuum. To have a more complete description of the microscopic properties of a black hole, general relativity is not enough; it must be combined with quantum mechanics. Quantum theory describes the Universe on a microscopic scale. It was formulated in the first decades of the 20th century, and its results challenged the beliefs that physicists had about nature.

As a first step, we need to remember the aforementioned Heisenberg uncertainty principle. This principle requires that some physical quantities, such as energy and time or position and momentum (the latter is the product of mass and velocity), are subject to a certain degree of uncertainty. In other words, in a physical system, the values of physical observables, such as those mentioned above, cannot be determined simultaneously, unlike what happens in classical physics, where all the observable quantities are well defined and hence describable by their values. According to Heisenberg's uncertainty principle, the more precisely the position is measured, the less we know about the moment, and the tyranny of the uncertainty principle also applies to energy, E, and time, t. The important thing to remember is that this principle does not depend on the precision of the experiments. Even in an ideal and perfect experiment, indeterminacy would continue to exist.

So, not only it is not possible to know the exact energy of a system at a certain instant, but if we wanted to exactly determine the energy of the system, we would need infinite time. As a consequence, a system does not have a defined energy at every instant, but it changes, or fluctuates, permanently, that is, its value increases and decreases compared to what we would expect, and all this at such a speed that it cannot be measured directly.

In other words, *energy conservation* can be violated, but only for very short periods of time: the less energy there is in the fluctuation, the longer it can persist. Quantum uncertainty allows small amounts of energy to appear from nothing, always on the condition that they disappear in a very short time. This energy can take the form of very short-lived pairs of particles and antiparticles, or *virtual particles*, for example an electron–positron pair. Therefore, Heisenberg's uncertainty principle means that the most uniform environment necessarily presents irregularities at the quantum level. From the point of view of quantum mechanics, the vacuum is a sort of sea full of particles and antiparticles (virtual particles), which are born and annihilate rapidly and which ensure that the energy of the field cannot be completely zero. If these vacuum fluctuations occur near the

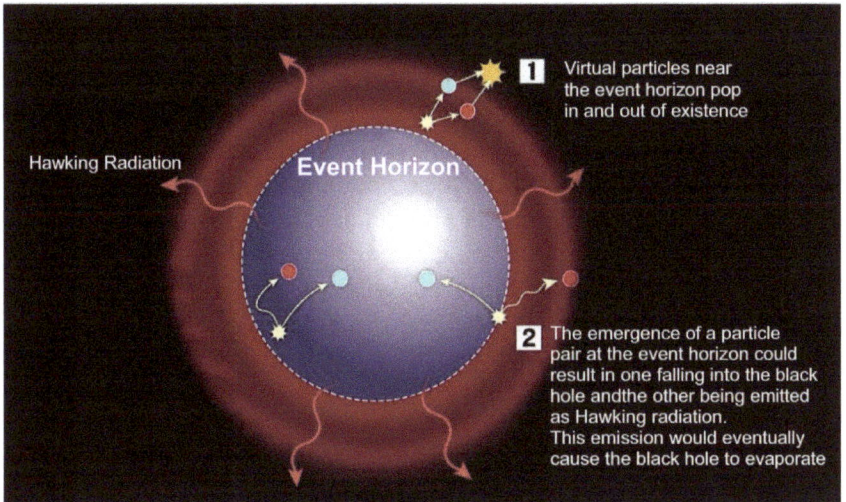

Figure 5.4.   Evaporation of black holes.

event horizon, given that the gravitational attraction is fierce, it is possible for one particle to fall into the black hole before being reabsorbed by the vacuum and the other to escape from it. Observing from the outside, it is as if the black hole emitted energy in the form of particles (Figure 5.4). Consequently, for the conservation of energy, the particle that escapes carries with it positive energy, and the one that falls into the black hole carries with it negative energy. The black hole gives energy to the particles; therefore, it loses mass, and over time, it would disappear. This process, never observed experimentally, is called the *evaporation* of the black hole, and the energy emitted is responsible for the temperature, i.e. *Hawking radiation*.

The black hole gets smaller and hotter. The evaporation process is very slow and grows with the cube of the black hole's mass. In the case of a solar-mass black hole, the time required for evaporation would be of the order of 10^{67} years. A much less massive black hole, with a mass of 1.7×10^{11} kg, would take an evaporation time equal to the age of the Universe. Now, cosmology predicts the existence of black holes that formed immediately after the big bang with small masses. They are called *primordial black holes* and are produced by local variations in the density of the early Universe. Cosmology predicts that at 10^{-43} s, primordial black holes had a mass of 10^{-5} g. After 1 s, the mass would be 10^5 solar masses.

Black holes with a mass of 10^{-5} g would take 10^{-48} years to evaporate, i.e. there would be instantaneous evaporation. For masses equal to 1.6×10^{11} kg, the evaporation time would be 10,000 million years; therefore, they would be evaporating gradually. During evaporation, they should produce jets of particles similar to those in accelerators. In order to find signs of evaporation, gamma-ray telescopes have been used in space, but attempts to find these evaporating primordial black holes have led to nothing. Another possibility is to create mini black holes inside the Large Hadron Collider in Geneva. These would decay very quickly, proving that evaporation is a real effect. However, as of yet, there have been no positive results from this point of view either.

The Information Paradox

Given that, as Wheeler had established, black holes "have no hair", that is, given that the final state of a black hole does not depend on how the collapse occurred, that it could have been formed from infinite configurations, and that we can only know mass, charge, and angular momentum, this creates a problem. In short, a black hole should contain information which, however, cannot be accessed. Information requires energy, and there is a limit to the amount that can be stacked in a certain region. According to Einstein's famous relation, $E = mc^2$, which establishes the equality between energy and mass, mass is associated with information, and if a region contains a lot of information, it can collapse to form a black hole. The size of the latter must reflect the amount of information that generated it. We have seen that black holes evaporate, lose mass, and then disappear with a large release of energy. The question arises as to whether it is possible, after this happens, to recover the information that was contained in it. For quantum mechanics, according to which information cannot be destroyed, this must be possible. There are, however, two problems. The first is that all black holes of the same mass, charge, and angular momentum are indistinguishable. The second problem is linked to the nature of the evaporation of black holes: Hawking radiation is thermal and therefore random. Consequently, we will not be able to know which black hole it comes from. So, it would seem that black holes, in contradiction to quantum mechanics, destroy information. Many studies have attempted to resolve this paradox. One group of researchers argued that quantum mechanics must be modified, while another tried to find mechanisms to recover information from Hawking radiation. Gerard 't Hooft and Leonard Susskind proposed a solution related to *string theory*, according to which the particles would be produced by oscillations of microscopic strings. According to the idea of 't Hooft and Susskind, black holes could conserve information by altering the event horizon during the process of emitting Hawking radiation. The gravitational field of the particles falling into the black hole could deform the event horizon, which could modify the particles that exit the black hole, leaving a sort of "signal" on them that could allow us to recover information about the particles themselves. Recently, in 2013, Juan Maldacena and Susskind proposed another solution, proposing that the incoming and outgoing particles are connected to each other by a space-time tunnel and that this connection allows the information to be recovered.

Our discussion has highlighted that black holes are truly among the strangest and most mysterious objects in the Universe. They are laboratories to verify the laws of physics that we know and, at the same time, in some cases, question them. As we have seen, getting too close is very dangerous, so it is better to stay at a safe distance.

Chapter 6

Can We Travel Through Time?

*Time is the most undefinable yet paradoxical of things; the
past is gone, the future is not come, and the present becomes the
past even while we attempt to define it, and, like the flash of
lightning, at once exists and expires.*

— Charles Caleb Colton

The idea of time travel is very fascinating, but at the same time, ordinary people consider this possibility as the fruit of fantasy typical of science fiction. In fact, from past centuries to the present day, many science fiction novels are based on the idea of time travel. Mark Twain, in his novel *An American in King Arthur's Court* (1889) imagined that the protagonist of the novel, Hank Morgan, found himself in the time of King Arthur. In the novel *The Time Machine*, H. G. Wells imagined that time was the fourth dimension and that it was possible to move back and forth on this dimension as on spatial dimensions. For ordinary people, the idea of time travel seems like something so imaginative and full of paradoxes that it is not feasible. Many physicists and even science fiction writers are of the same opinion. Lester del Rey wrote in 1979 that time travel is impossible, and writer Orson Scott Card, in 1990, argued that time machines were magic tricks used by science fiction writers. Aside from the problem of devising a technology that would allow a time machine to be made possible, one of the big problems with time travel is the logical paradoxes it entails. One of the most famous is the *grandfather paradox*, according to which if someone could go back in time by killing their grandfather, he could not be born; therefore, in reality, he would never have existed. So, if time travel were truly possible, there should be laws regulating such travel. For example, in the mid-1980s, Russian physicist Igor Dmitriyevich Novikov introduced the Novikov *self-consistency principle*. According to this principle, it is not possible to change the past. In a situation where one can travel to the past, events are determined not only by past ones but also by future ones; the laws of physics only allow time travel, which does not create paradoxes. It is impossible to prevent an event that has already occurred from occurring in the future. The Novikov principle was generalised by Allen Everett and Thomas Roman, defining the so-called *banana peel*, or *non-interference mechanism*. That is, if someone travels to the past to kill his grandfather, something will happen to prevent him. Stephen Hawking went so far as to claim that time travel is impossible. In 1991, he stated the chronology protection conjecture, that is, physical laws prevent time travel. Another idea for resolving the grandfather paradox is based on the *many-worlds interpretation of quantum mechanics*. Applying this theory to the case of the grandfather paradox, things would go as follows. If you return to the past to kill your grandfather, you will enter a parallel world distinct from the previous one, that is, a world in which you exist only as a time traveller, not in relation to the grandson of the person you want to kill. In this way, the paradox ceases to exist. With current scientific knowledge, we can answer the question: Can we travel through time? The key to the answer lies in Einstein's theory of relativity. If we talk about travel into the future, the answer is yes, while if we think about travel into the past, giving an answer is more difficult. In any case, we cannot categorically exclude them. We will see that travel into the past could be possible thanks to the curvature of space-time, which would allow the construction of a time "machine". In short, to be able to travel into the future, we would need to be able to "follow" a ray of light, that is, move at speeds close to that of light. Conceptually, it is simple, but with current technologies, travelling at speeds comparable to that of light is not possible. Instead, to travel into the past, you need to "anticipate" a ray of light.

This could be done in two different ways: by using shortcuts in space-time, as we will describe later, or by moving faster than the speed of light. The theory of special relativity, as we have seen, does not allow a massive object to reach or exceed the speed of light. However, the theory does not prohibit the existence of particles that always move at speeds greater than that of light, without being able to go below this speed. Such particles are called *tachyons*. The first theoretical description was due to Arnold

Sommerfeld, followed by attempts at interpretation within special relativity by George Sudarshan in 1962. Gerald Feinberg coined the term "tachyon" and popularised the theory in the 1960s. Curiously, unlike ordinary particles, the speed of a tachyon increases as its energy decreases. For tachyons, time does not stand still but actually flows backwards: in other words, tachyons travel backwards in time. Today, there are serious doubts about the existence of tachyons, and even if they existed, they could not be revealed due to problems in their interactions with ordinary matter. Assuming that tachyons exist and could interact with ordinary matter, there could be a violation of causality, the principle according to which the cause always precedes the effect. There would no longer be any way to distinguish the difference between the future and the past. A particle could send energy or information into its past. If we could use these particles or, paradoxically, exceed the speed of light, we could send ourselves the results of the football pools the day before playing and be sure of winning. This way of travelling into the past by "anticipating" a ray of light is not feasible, but as mentioned it could be done by taking a shortcut in space-time. The gravitational effect of very massive and dense objects, such as neutron stars or black holes, could distort space-time to the point of producing closed time curves as well as paths that take us into the past.

Journey Into the Future

Journeys into the future are depicted in some films, such as *Planet of the Apes*, in which four astronauts head towards Alpha Centauri, starting from Earth. The journey lasts a few years, and suddenly the spaceship crashes onto an unknown planet. At the end of the film, the protagonist realises that the planet is Earth but two thousand years in the future, during which apes have evolved and conquered the planet. This film, although science fiction, exactly describes the mode of such a journey into the future. The first to realise this possibility was Paul Langevin in 1911. According to him, to travel into the future, it is enough to accelerate to high speeds and then return to Earth. If the spacecraft were to reach speeds close to that of light, the journey will consist of a long acceleration phase, a period of constant speed, and then a long deceleration phase to return to Earth. The elapsed time for the astronaut, who is in motion, as we saw in Chapter 3, is the proper time, which is always less than the time for someone who is stationary because the distance travelled by the clock is subtracted from the observer's time. To calculate proper time using special relativity, you

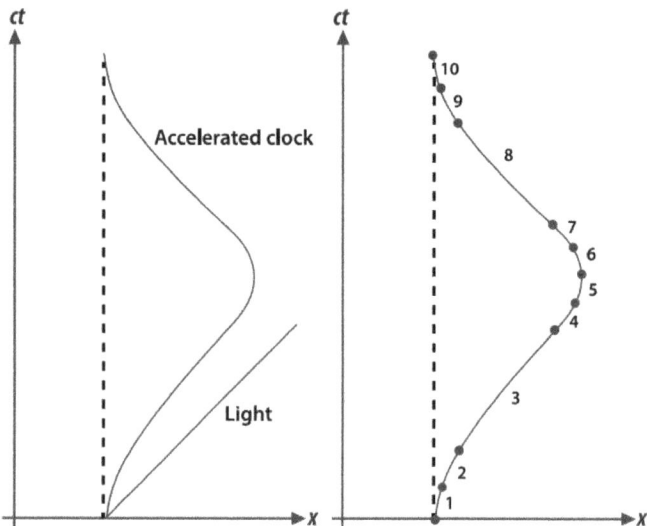

Figure 6.1.   Left: Worldline of an accelerated clock moving away from a point and then returning to it. Right: The left line approximated by 10 straight segments. One segment indicates that the speed is constant and the clock is inertial.

need to divide the journey in space-time into sub-paths in which the velocity is constant and use the formula for time dilation in special relativity (see Chapter 3). For a spaceship that accelerates for a certain time, reaches cruising speed, and then decelerates to return to the starting point, the Minkowski space-time diagram is given by Figure 6.1. In the Minkowski diagram, a straight line segment indicates that the speed is constant. In Figure 6.1 (left), you can see that the curve is not a straight line, and there is a variation in the slope of the curve at each point. To apply the equations of special relativity to calculate time dilation, the curve must be divided into as many segments as straight as possible. This is done in Figure 6.1 (right). The curve is divided into 10 approximately straight parts. In each segment, the clock is approximately inertial, and the rules of special relativity apply to calculate proper time. In each segment, the proper time is given by $\tau^2 = t^2 - x^2$. The proper time (the astronaut's time) is less than the elapsed time for the inertial observer because the square of the distance travelled by the clock is subtracted from the time t. The total proper time is the sum of all segments, i.e. of all proper times, and is less than the time measured by the inertial observer.

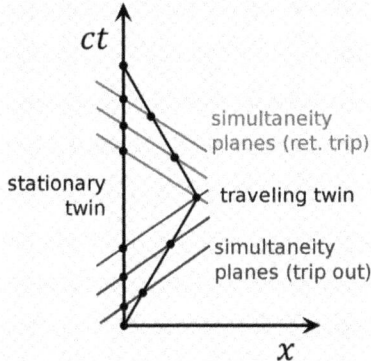

Figure 6.2.   Alberto's worldline, denoted by triangle in black.

Let us now consider two twins, Francesca and Alberto. Alberto makes a journey from Earth on a spaceship at a speed equal to 0.995 the speed of light and then returns to Earth after a year, a time measured with his watch. What is the elapsed time for Francesca? To do the calculation exactly, we should use the line in the Minkowski space in Figure 6.1, which shows how Francesco accelerates between points 1 and 3, decelerates to reverse course between points 4 and 5, accelerates again between 5 and 8, and finally decelerates between points 9 and 10. As we said before, to find Alberto's proper time, we would need to add all these segments.

To simplify the calculation, we neglect the acceleration and deceleration periods. So, let's suppose that Alberto starts directly at a speed of 0.995 times the speed of light; once he reaches the halfway point, he instantly reverses his speed to reverse course and continues moving towards the Earth, always at the same speed. In Minkowski space-time, the line on which Alberto will move is the black triangle shown in Figure 6.2.

In this way, we can use the time dilation rule to calculate the proper time that has elapsed for Francesca. From Chapter 3, we know that $\Delta t_{\text{Francesca}} = \Delta t_{\text{Alberto}}/\sqrt{(1 - v^2/c^2)} = 1$ year$/\sqrt{0.995^2}$, which is approximately 10 years. In this way, Alberto finds himself in Francesca's future. This is an example of a journey into the future. This topic, known as the *twin paradox*, was the subject of great debate in the first half of the 20th century due to a misinterpretation of special relativity. The paradox consisted in the fact that if all reference systems were equivalent, Alberto, in his reference system, would see that it was Francesca who moved and made the journey; therefore, he should be the one who aged. Herbert Dingle, an English philosopher and physicist, thought he could prove the invalidity of special relativity using this argument, but Einstein and Bohr showed that his reasoning was incorrect. The paradox arises when we do not take into account the fact that the reference systems of Francesca and Alberto are different, and that proper time is invariant only for inertial systems. Francesca's system is inertial, while Alberto's is not and is subject to accelerations and decelerations. So, Alberto cannot calculate Francesca's proper time using special relativity, which deals with constant velocity, or inertial, systems. The presence of accelerations and decelerations makes the principle of relativity invalid, and it is clear that it is Alberto who moves and not Francesca. So, there is no paradox. On the other hand, there are experimental demonstrations of the twin paradox, as we saw in Chapter 4, such as the experiment by Joseph Hafele and Richard Keating in 1971, who compared the times of an atomic clock placed on a plane flying around the world and a twin watch left on the ground. As already mentioned, the clock on the plane had lagged behind the one on the ground. We gave another example when we talked about the muons of cosmic radiation in Chapter 3, which, due to their high-speed motions, have a lifetime longer than what would be expected if they were stationary. In the experiments carried out in the Large Hadron Collider (LHC), protons reach speeds very close to that of light (0.999999991 of the speed of light) and collide to generate radioactive particles, such as muons and pions, which travel at speeds close to that of light. The proper time of particles moving at high speeds is less than that spent in the laboratory; therefore, these particles have travelled into the future.

Having established that it is possible to make journeys into the future, we can now ask ourselves whether we can make such a journey with the means at our disposal. If we want to be precise, every time we move, we take a journey into the future, but we move into the future by an infinitesimal amount.

For example, the Russian cosmonaut Sergei Krikalev, remaining in orbit for about eight hundred days at a speed of 27,000 km/h, made a journey into the future of about 0.02 seconds. For example, if we were on the Juno probe which barely reaches 300 km/s, or 0.1% of the speed of light, for a year, with Jupiter as the destination, we would have made a journey into the future of almost 16 s – very little indeed. We understand that real travel into the future requires great speed. If we could move in a spacecraft at 10% of the speed of light for one year, 1,005 years would have passed on

Earth, or less than two days. If we travelled at 80% of the speed of light for a year, one year and eight months would have passed on Earth. If we want to travel for significant periods in the future, even higher speeds are needed. At 99.99% the speed of light, for one of our years, almost 71 years would pass on Earth and travelling at 99.999% the speed of light, for one of our years, almost 224 years would pass on Earth. These speeds are absolutely out of our reach. As we saw in Chapter 3 (Figure 3.5), the energy required to approach the speed of light grows exponentially. We had given the example of a 10 ton spacecraft. To accelerate it beyond 70% of the speed of light would require the energy produced by the European Union in 133 years! If we tried to push it to 99% of the speed of light, we would need about five times more energy, equal to the energy produced by the European Union in 665 years. In other words, although travel into the future is conceptually simple, there are technological limitations. The energy provided by fossil fuels is not enough. A nuclear fusion engine could be used, but we have not yet developed fusion reactions on Earth. There remains the possibility of using antimatter. An antimatter particle has the same characteristics as a matter particle but has the opposite charge. When brought into contact with a particle of matter, the two annihilate, emitting energy. If we wanted to reach 99.99% of the speed of light, we would need 350 tons of antimatter. Unfortunately, antimatter is very difficult to produce and has exorbitant costs. According to some estimates, US$25 billion would be needed to produce 1 g of positrons, and according to NASA, US$62,500 billion is needed to produce 1 g of antihydrogen. In addition to the problems of antimatter production, there are problems in the exploitation and control of the energy coming from the annihilation of matter and antimatter. Other proposed ideas are the use of huge sails pushed by the solar wind or by powerful lasers placed on Earth – all science fiction ideas for the moment. In addition to energy problems, we also have other, even more practical problems. A spacecraft moving at close to the speed of light would risk being destroyed by collisions with micrometeorites. The gas found in space colliding with the spacecraft would produce jets of radiation very similar to cosmic rays; and it would therefore be necessary to shield the spacecraft from these lethal radiations. In short, it seems that nature does not want us to reach speeds close to that of light to travel into the future. There is another way to travel into the future using general relativity. In Chapter 4, we saw that time slows down in the presence of gravitational fields. Near a black hole, time passes very slowly. Time travel of this type is described in the film *Interstellar*. The concepts depicted in the film are very realistic, as physicist Kip Thorne made the calculations so that they are mathematically correct. In one part of the film, the *Endurance* spacecraft approaches a supergiant, rotating black hole called *Gargantua* with a mass equal to 100 million times that of the Sun, a mass intermediate between that of the black hole in our galaxy and that of M87. A planet orbits near the event horizon, onto which the astronauts descend, while the spacecraft remains far away. From Thorne's calculations, one hour on the planet corresponds to seven years on the spacecraft. When the astronauts who landed on the planet return to the spacecraft, after a few hours, they see that the astronauts on the spacecraft have aged 20 years. This situation is similar to what happens in the *twin paradox*; the difference is that the slowing down of time is not due to motion but to the intensity of the gravitational field. This possibility of travelling into the future is also not trivial. The supermassive black hole, of about 4 million solar masses, closest to us is that of our galaxy and is located about 28,000 light years away. The closest to us is Gaia BH1 at 1,600 light years and having a mass of 10 solar masses, which is also too far from us and it is possible that black holes exists in the Iades, a group of stars, at 150 light years.

Journey Into the Past

Travelling into the future, as we have seen, is achievable as long as we are talking about small intervals of time. If we want to make leaps into the future years, decades, or centuries ahead, we need to move at speeds close to that of light, and we have seen that this is not currently within the reach of our technology. The question remains as to whether it is possible to travel into the past. Let us first consider the case of special relativity, in which space-time is flat. In this case, to return to the past, it would be necessary to move on a worldline like the one drawn in Figure 6.3.

The figure shows the worldline of a particle starting from *A*, moving in space-time towards the future until the line curves, and begins to descend. From the curve up to *B*, the particle moves towards the past. However, the part in which the particle moves towards the past is not allowed because, as we saw in Chapter 3, if in a Minkowski space-time diagram the inclination of the line is greater than 45°, the particle would travel at a speed greater than that of the light, and this is not possible. So, if we want to travel into the past, we need lines of closed universes but which, in some way, do not have the problem of the line in Figure 6.3, where, as mentioned, in the stretch after the curve up to *B*, the speed is

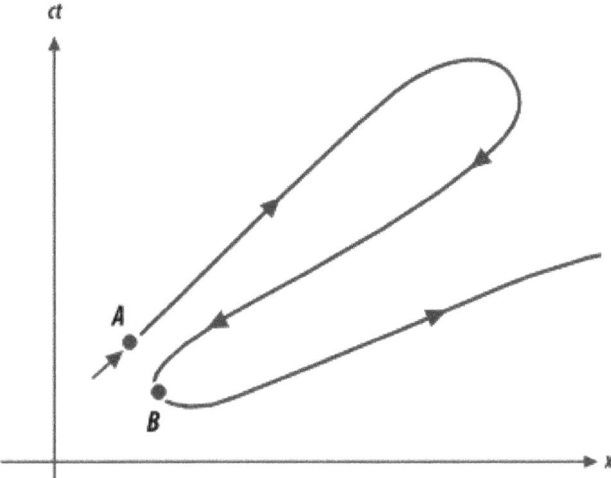

Figure 6.3.   A forbidden worldline in a flat space-time.

greater than that of light. This makes us understand that in a flat space-time similar to that of special relativity, one cannot travel through time. We need a curved space-time. To understand how we can create a *closed worldline* without particles moving faster than the speed of light, we begin by introducing the so-called *cone of light*, shown in Figure 6.4. We represent our present as a point, *O*. We send rays of light from *O* in all directions.

The result is a cone, as shown in Figure 6.4, which, as mentioned, is called a *cone of light*. The entire upper cone, starting with *O*, is the cone of light of future events. A particle with mass will move within this cone. The *worldlines* cannot exit the cone because that would imply that we are moving faster than the speed of light. The lower cone contains all past events. If an event in the past can influence an event in the future, we will say that the two events are causally linked, i.e. the principle of causality, or the cause–effect principle, applies. In other words, the effect always follows the cause, as we are used to it. According to the principle of causality, our present, i.e. point *O*, can only influence what will happen in the future light cone and can only be influenced by what happened in the past light cone. Every point in space-time has its own cone of light. As we mentioned, in a flat space-time, travel into the past is not possible because the future of *O* is always found in the future light cone, which is separated

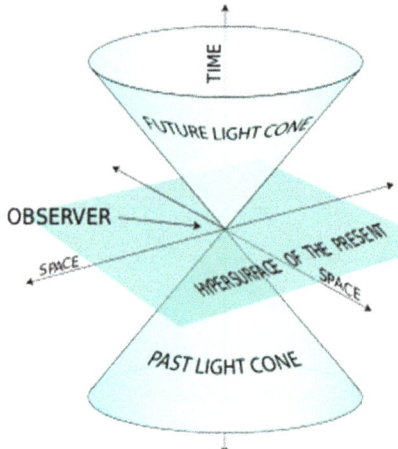

Figure 6.4.   Cone of light.

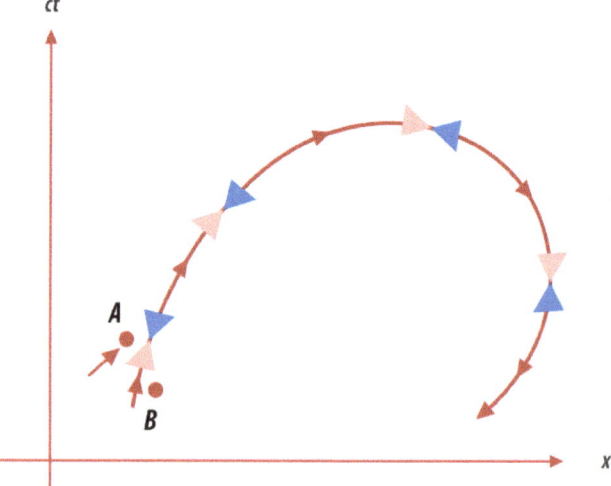

Figure 6.5.   Worldline in a curved space-time.

from the past light cone. However, if we are in a curved space-time, the orientation of the light cones changes due to the presence of gravity.

At each point in space-time, a cone of light can be drawn. The orientation of the light cone changes due to the curvature of space-time. Therefore, there can be worldlines that curve in the direction of past time while maintaining a speed lower than that of light, as seen in Figure 6.5. As can be

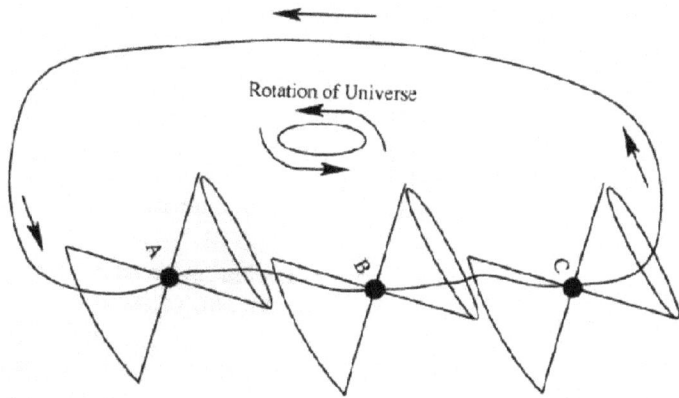

Figure 6.6. Closed worldline in a rotating universe. The light cones are tilted. For very fast motion, the future of *A* is located in the past of *B*, and the future of *B* in the past of *C*. Ultimately, the future of *A* is located in its past.

seen, the worldline is maintained in every light cone of the future, which implies that the speed of light is not exceeded, and locally, the particle obeys the laws of special relativity. Looking at the worldline globally, you see that it bends in the direction of past time. Ultimately, to build a time machine, you need a distribution of matter that allows closed worldlines and, at the same time, is a solution to the equations of general relativity.

The famous mathematician Kurt Gödel, of German origin but born in the Czech Republic, was one of the first to show the possibility of travelling to the past by finding solutions to general relativity. In 1949, he published a paper about a rotating universe, satisfying the equations of general relativity. In this universe, there are naturally closed worldlines that allow travel into the past. The initial idea that pushed Gödel to find these solutions was to solve the problem of finite universes that tend to collapse. In Gödel's universe, the centrifugal force produced by rotation avoided collapse by counteracting gravitational attraction. The rotation of masses within the universe produces a distortion in space-time, resulting in the tilting of light cones, as shown in Figure 6.6. A traveller starting from *A* will move into his future light cone, through the cones of light of *B* and *C*, and then return to the past light cone of *A*. As you can see, the traveller's *worldline* is always within its local light cone and therefore will not exceed the speed of light. According to Gödel's calculations, his universe would take 70 billion years for one rotation. Obviously, Gödel immediately realised the paradoxes introduced by the possibility of time travel. At the beginning of this chapter, we talked about the *grandfather paradox* and the violation of the *principle of causality*, which is, as already mentioned, the principle according to which the effect always follows the cause. Gödel moved to the United States in 1939 and taught at the Institute for Advanced Study in Princeton. Einstein worked at the same institute, and the two walked together every day. So, Einstein knew about Gödel's universe and realised the paradoxes it implied.

In a comment on Gödel's article, he wondered whether the paradoxes produced by travelling into the past could not be excluded on the basis of physical principles. In 2020, researchers from Kansas State University, Lion Shamir and collaborators, argued that the primordial Universe may have been equipped with rotation, but observations of the current Universe exclude it from rotating; therefore, there are no paths that can take us back to the past. So, Einstein's and Gödel's concerns had no reason to exist. An early example of a time machine was described by Willem Jacob von Stockum and Kornel Lanczos in 1924. It was a long column of dust kept in rotation. However, only in 1974, an analysis by Frank Tipler, a student who was working on his doctoral thesis at the University of Maryland, made it possible to understand that a rapidly rotating cylinder can produce the inclination of light cones and create closed worldlines. By moving around the cylinder at very high speeds, but less than the speed of light, one can travel into the past. Tipler, in his work, was interested in the construction of a time machine and had published some articles in which he answered the question of whether it was possible to create such a machine with the use of materials consisting of ordinary matter. According to his studies, this was not feasible because such a machine would lead to the formation of singularities, i.e. black holes. To avoid this problem, it would be necessary to use a particular material with negative mass and energy density that would have a repulsion effect on normal matter. This material is referred to as *exotic matter*. Ultimately, a Tipler time machine would be theoretically feasible, but the practical difficulties would be enormous or insurmountable.

One diagram of the Tipler cylinder is drawn in Figure 6.7. A cylinder of mass approximately 10 times that of the Sun rotates rapidly around the time axis. The rapid rotation deforms space-time as well as prevents the structure from collapsing. As can be seen from the figure, the light cones far from the cylinder are unaffected and point upwards, and as they approach it, they tilt, with the future light cone directed in the direction of rotation. To travel through time, a traveller with his spacecraft would have

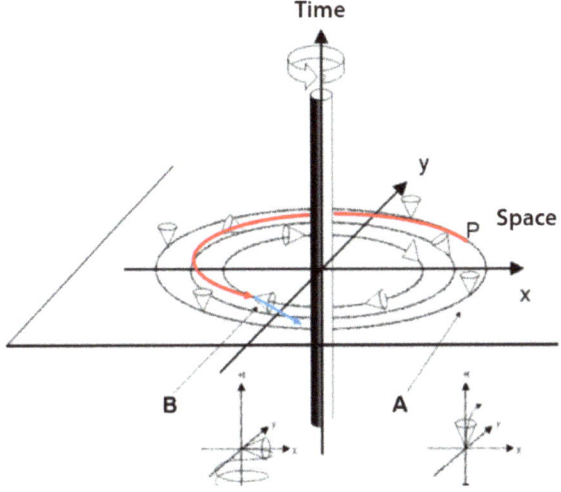

Figure 6.7. Tipler cylinder.

Source: Reproduced from http://www.infinitoteatrodelcosmo.it/2017/01/02/viaggiare-nel-tempo-basta-un-cilindro/

to head towards the cylinder where space-time is distorted. Seen by an observer on Earth, the traveller *P* spirals towards the cylinder (red curve) in the space defined by the *x*- and *y*-axes along the negative direction of time. As it gets closer to the rotating cone, the cone of light bends more and more towards it, tilting more than 45 degrees. Trajectories at speeds slower than that of light are possible, so the traveller can easily go back (blue line) compared to the original time (when he was in *A*), given that his time axis has become *x*. More simply, the traveller approaches the cylinder and navigates along a helical trajectory, moving towards the negative part of time while moving along trajectories with speeds lower than that of light. After reaching the desired past time, he simply moves away from the cylinder, exits the propeller, and then heads towards Earth. Tipler calculated the rotation speed of the cylinder, which should be one rotation every millisecond; its mass, as mentioned, should be equal to 10 times that of the Sun, with a radius of 10 km and a length of 100 km. In nature, there are pulsars, rotating neutron stars, with speeds compatible with that of a Tipler cylinder with the same mass and radius. So, it would be necessary to be able to bring 10 neutron stars to a certain position and stack them on top of each other to form the cylinder and make them rotate at the same speed and in the same direction. Finally, exotic matter would have to be used to prevent such a massive system from collapsing into a black hole. It is immediately clear that creating such an object is most likely impossible, certainly not with the current technology. Another way to travel through time without some of the problems of Tipler's method, such as avoiding the appearance of black holes and the need to use exotic matter, was invented in 1990 by Richard Gott III of Princeton University. The method is based on *cosmic strings*, as suggested in 1976 by physicist Tom Kibble. These objects, never observed until now, would have formed in the pri-mordial Universe and are sort of very long filaments, reaching up to the size of the entire Universe, and are very thin, with a thickness less than that of an atomic nucleus. However, they are very massive. One centimetre should have a mass of a few million billion tons. Despite their large mass, the strings do not attract each other, as they are made up of non-ordinary matter with negative pressure that compensates for the gravitational attrac-tion. Due to their large mass, space-time at these filaments curves consid-erably. The string produces a curvature of space-time similar to that of a cone, causing the space in their vicinity to shrink. Richard Gott solved Einstein's equations by considering two cosmic strings in motion, which however never intersect. From the solution, it turned out that by making them move rapidly apart, closed time curves were produced around the strings that allowed time travel. To be able to travel through time, it is necessary for a "chrononaut" to move at a speed close to that of light in the vicinity of the strings, where the closed time curves are located. Using the ideas of his time machine, Gott went further and argued that we are living in a universe that is its own mother. In Gott's words:

In 1998, Li Xin-Li and I discovered that a time loop would allow the cosmos to generate itself. In this model, the cosmos creates itself in a sort of closed circuit, so we return to the starting point and start again

Another way of travelling through time is through the so-called wormholes, a term coined in 1957 by John Archibld Wheeler. Wormholes were discovered by Ludwig Flamm in 1916 as solutions to general relativ-ity. In 1935, Einstein and his collaborator Nathan Rosen, unaware of Flamm's study, rediscovered them and studied their properties. These solutions to general relativity were called Einstein–Rosen bridges and renamed wormholes by Wheeler. These are tunnels in space-time.

In Figure 6.8, our Universe is represented by a folded sheet structure. A cut in hyperspace connects points *A* and *B* with the wormhole. The

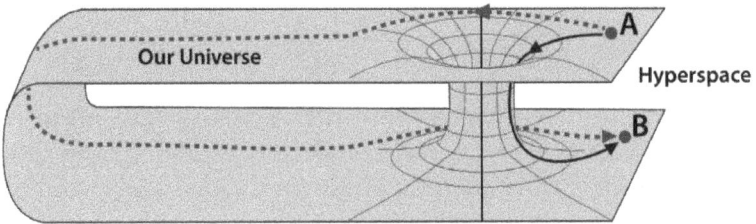

Figure 6.8.   A wormhole.

tunnel is characterised by two mouths, which are spherical in three-dimensional space connected to each other by the tunnel, which runs in *hyperspace*. The term hyperspace was introduced in 1867 by Arthur Cayley and designates a space having a number of dimensions greater than three. For example, space-time is a four-dimensional hyperspace. The first theory of hyperspace, or extra dimensions, was proposed by the Polish mathematician Theodor Kaluza in 1919. He added a fifth dimension to the four dimensions of space-time to be able to introduce electromagnetism into the four-dimensional theory of general relativity. In 1926, Oskar Klein hypothesised that the fifth dimension was not visible because it was curled up to form a tiny circle of 10^{-33} cm. Today, there are various theories based on the existence of extra dimensions, such as superstring theory. This theory is based on the idea that particles are tiny vibrating strings in 10-dimensional space-time. To get from point A to point B in Figure 6.8, there are two possibilities: move through the Universe, or take a much shorter path through the tunnel. After the discovery of wormholes, no one imagined that they could be used to move quickly through hyperspace. A wormhole is created by two black holes when the singularity of one of them reaches the singularity of the other, and a tunnel is created. As was shown by Wheeler and Robert Fuller, such structures are unstable, however. They are created quickly and destroyed just as quickly. The throat of the tunnel can collapse so quickly that no light can pass through the wormhole. In the late 1980s, Kip Thorne and his students, Mike Morris and Ulvi Yurtsever, published papers on traversable wormholes. In 1988, Thorne published a famous paper in which he used general relativity and quantum mechanics to speculate if it was possible to open and keep open a wormhole using exotic matter. The idea of using wormholes as time machines came to Thorne's mind when Carl Sagan, who was writing the novel *Contact* and had to instantly land the novel's protagonist on the star Vega, 26 light years away, asked Thorne if there were ways of doing it. Thorne had the idea of using a wormhole that would not close thanks to the use of exotic matter – matter with negative energy. Two questions arise: the first is whether wormholes exist in nature, and the second is whether negative-energy matter exists to keep them open. According to John Wheeler, wormholes exist naturally in the primordial phases of the Universe in the form of *quantum foam*. The space-time of the primordial Universe would be made up of a network of space-time tunnels that continually form and disappear due to the uncertainty principle of quantum mechanics. The size of these wormholes would be of the order of 10^{-33} cm. Quantum foam and therefore primordial wormholes are speculations. To be sure of their existence, we need to have a theory, still undeveloped, called quantum gravity, which combines general relativity and quantum mechanics and would allow the study of space-time in the early days of the Universe. If quantum foam existed, negative energy would be needed to widen the wormhole and stabilise it. Negative energy really exists and is linked to an effect called the Casimir effect. As described in Chapter 8, space is never completely empty. Quantum uncertainty allows small amounts of energy to appear out of nothing, and this energy can take the form of pairs of particles and antiparticles, virtual particles, or virtual photons. In 1948, Hendrik Casimir showed that two parallel metal plates in a vacuum, separated by a small distance, experience an attractive force (Figure 6.9). The force arises because the structure of the vacuum is modified by the presence of the conductive plates. Simply put, the plates are contained in a vacuum. The wavelength of the virtual photons outside the plates is not limited by their presence, while the virtual particles inside the plates must have a wavelength that is an integer submultiple of the distance between the plates.

Therefore, more fluctuations occur outside the two plates than between the plates, tending to bring them closer together. Ultimately, the number of photons inside the plates is lower than the number of photons outside. This means that, on average, there is less energy between the plates than in a vacuum. Since the energy in a vacuum is the minimum possible – let's say it is zero – the energy between the plates is negative. So, a wormhole could be held open, as Thorne proposed, by lining its mouths with pairs of metal plates. From what we've seen, a wormhole is a structure that allows you to travel from one place in space-time to another almost instantaneously. We need to understand how this feature of wormholes can be used to travel through time. The mouths of the

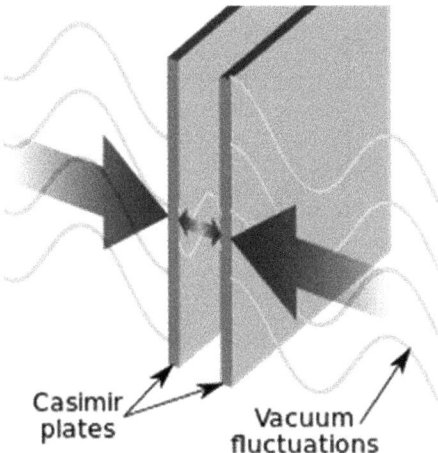

Figure 6.9.   Casimir effect. Only fluctuations with a wavelength equal to the distance between the plates are permitted inside the plates. Outside, the wavelength of the virtual particles does not have this limit, and this produces a force that tends to bring the plates together.

Source: Wikipedia, Davide Mauro.

wormhole represent the entrance to the path that will be taken in hyperspace, or space-time that allows us to overcome a ray of light. Looking at Figure 6.8, we can see that if a ray of light moves from *A* to *B* through our Universe, it will take longer than that spent passing through the wormhole from *A* to *B*. That is, by entering one mouth of the wormhole and exiting the other, we will have passed a ray of light that follows the indicated path in space. After Thorne showed Sagan how to reach two distant regions of space through a wormhole, his colleagues pointed out that wormholes could connect different times since they are in four-dimensional spacetime. To visualise the idea, let's consider a three-dimensional Minkowski space with two spatial dimensions and a temporal one (Figure 6.10). In Figure 6.10, time flows upwards, and you can see that the wormhole has two mouths positioned on two sections at different heights, i.e. times. The mouths of the wormhole are located in hyperspace, outside the three-dimensional space of the drawing. If our "present" is in the BEFORE, moving through the wormhole, we move towards the future. If our

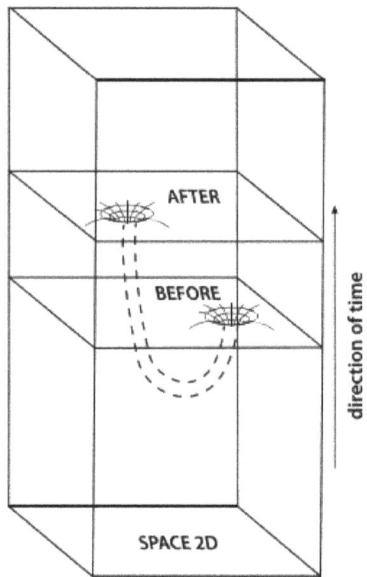

Figure 6.10.   Wormhole connecting two different times in a Minkowski space.
Source: *Black Holes, Wormholes and Time Machines*, Jim Al-Khalili.

"present" is in the AFTER, moving through the wormhole, we move towards the past. So, to travel through time, we need to:

- generate a "traversable wormhole", which can be achieved by taking one from quantum foam, inflating it to macroscopic dimensions, and stabilising it using exotic matter;
- place one of the wormhole mouths on a rocket;
- create a time difference between the mouths by making the rocket move at relativistic speeds or by making the mouth move around a black hole;
- bring the mouths closer together (so that by entering one and exiting the other, you travel through time), giving rise to a time machine.

The described process was studied by Thorne and collaborators. In 1988, Thorne, together with Mike Morris and Ulvi Yurtsever, published an article in the prestigious journal *Physical Review Letters* in which they showed how to obtain a time delay between the mouths of a single wormhole, thus describing how a wormhole could be used as a time machine. To describe their result, we use the space-time diagram in Figure 6.11.

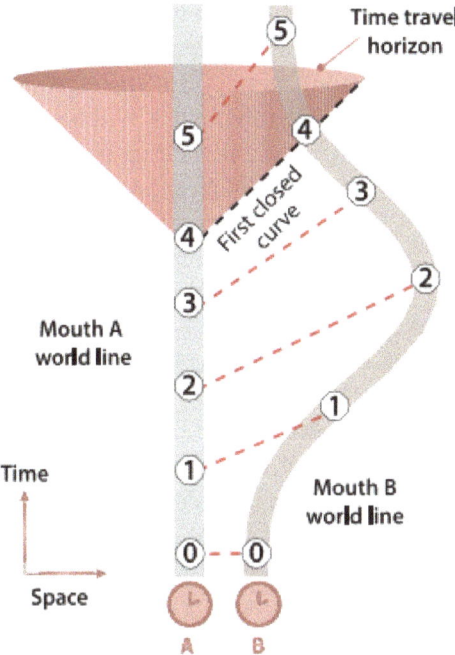

Figure 6.11.   Space-time diagram that allows you to convert a wormhole into a time machine.

Source: Adapted from Morris, Thorne, Yurtsever, 1988, *Physical Review Letters* 61, 13.

The mouth A of the wormhole is located on Earth, and its worldline is indicated by the vertical band, which essentially means that it does not move in space (it is located on Earth) but moves in time. Mouth B is located on a spaceship that starts from Earth and is sent into space until it reaches a speed close to that of light, and after 100 years, it returns to Earth. Suppose the length of the wormhole is small, say 1 m. Going from mouth B to mouth A takes you back 100 years in the past. How does this time jump work? As can be seen from Figure 6.11, the worldline followed by mouth B is a curve equal to that used in Figure 6.1, used to illustrate the *twin paradox*, which represented a spacecraft being accelerated to speeds close to that of light, then reversed course, and returned to Earth. As in the case of the twin paradox, the mouths of the wormhole are subject to time dilation, and by moving one at close to the speed of light and then returning it to the starting point, a time lag between the mouths is created. The circles connected by the dotted line indicate the points in space-time having the same proper times. In other words (using the same names as the twin paradox), if Alberto is on the spacecraft, and at point 2, his clock shows 2:00 pm, looking from mouth B of the wormhole, Francesca on Earth (at point 2 of the universe), that clock will also show 2:00 pm, and similarly for the other points. The line that joins point 4 is located in the cone of light and will have an inclination less than 45°, i.e. the speed is less than that of light. The lines connecting points 0, 1, 2, and 3 have inclinations greater than 45°. Since you cannot have closed universe lines with speeds greater than the speed of light, you can only have closed time lines starting from the point marked with 4. The line that joins point 4 is the first that demarcates the region of space-time in which travel into the past can occur.

The cone of light that encompasses this line shown in Figure 6.11 is called the *time travel horizon*, or *Cauchy horizon*. In the region of space-time within such a light cone, an observer can go to mouth B and move to A, returning to the past, even if it is not possible to return to subsequent events before the time machine was activated. An important thing to note is that you cannot go back to times before the time machine was activated. If it has been active since May 30, 2015, you will not be able to travel to an earlier date. After the 1988 paper was published, hundreds of other papers dealing with time travel using wormholes or otherwise were published in the 1990s. Not everyone agreed that time travel was possible. Some calculations have shown that the last phase, that of bringing the mouths closer together, could be disastrous for the wormhole. This is because when the wormhole is a time machine, the light that enters from a mouth and passes through it can return through normal space to the mouth from which it entered before entering, and the process would continue in a cyclical manner. A sort of cycle would be created that would amplify the energy moving in the wormhole until it explodes. Furthermore, quantum vacuum fluctuations can produce high-energy photons from nothing. Due to these fluctuations near the time travel horizon shown in Figure 6.11, infinite energy could accumulate, producing a black hole. However, these predictions are based on the *semiclassical theory of gravity*, an approximation to the theory of quantum gravity. In other words, we cannot be sure that things will go as planned. To be sure that there is some sort of law that does not allow the existence of time machines, you first need to have a complete theory of quantum gravity. The phenomenon of the destruction of the wormhole would be a sort of proof of the *chronology protection conjecture* enunciated by Stephen Hawking, i.e. the idea that physical laws

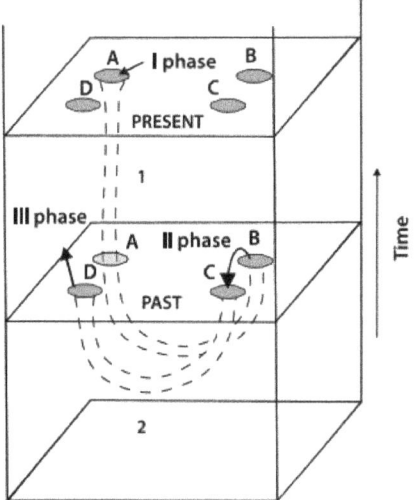

Figure 6.12.   Roman's time machine, using two wormholes.
Source: Black Holes, Wormholes and Time Machines, Jim Al-Khalili.

prevent time travel. In any case, to solve this possible problem, one could use two wormholes and build a *Roman time machine* (named after the physicist who proposed the method). Basically, the method eliminates the need to bring the mouths closer together, which, as we have seen, could lead to the destruction of the wormhole. In practice, after having created the temporal difference between the two mouths of the wormhole, distant from each other, a second wormhole could be used (whose mouths have no temporal difference) to bring its mouth closer to that of the first (Figure 6.12). So, you could go through the first wormhole, and use the second one as a shortcut to get back to the starting point, before the starting time. Referring to Figure 6.12, we start from mouth *A* and arrive at mouth *B* of the first wormhole.

Then, you exit *B* and enter mouth *C* of the second wormhole, near *B* of the first. Finally, you exit through mouth *D* of the second wormhole, which is next to mouth *A* of the first, in its past. By not bringing the mouths of the first wormhole closer, we would not have the problem discussed above. Another possible time machine using wormholes was proposed by Matt Visnier. In this case, the wormhole has a small time shift, so it does not function as a time machine.

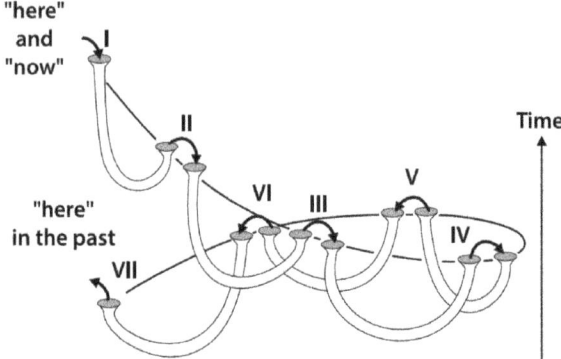

Figure 6.13.   Ring time machine.

Source: Black Holes, Wormholes and Time Machines, Jim Al-Khalili.

To obtain a ring time machine, you need to connect the wormholes, as in Figure 6.13. In this way, each wormhole allows a small time shift between the two mouths. The accumulation of time shifts adds up. By moving between the wormholes, you return to the starting point in the past. Here too, however, some studies have highlighted that a configuration of this type tends to resonate, with its possible destruction, again seen as an effect of the chronology protection conjecture. In conclusion, while relativity allows travel into the future and, at least on a theoretical level, travel into the past, to be sure that nature is not opposed to them, we need to wait until our knowledge of physics is more advanced.

Chapter 7

What is Dark Matter?

You know, dark matter matters.

— Neil deGrasse Tyson

The Beginnings

The stars of our galaxy, located on its disc, move in approximately circular orbits. Looking at the galaxy edge-on, the stars will move up and down like a rocking horse. A greater amount of local mass will produce a greater force of attraction on the stars and, therefore, a smaller amplitude in the stellar oscillations. Therefore, the study of stellar dynamics allows us to establish the local content. The problem of determining local density was first tackled by Lord Kelvin. Poincaré, from Kelvin's results, concluded that the quantity of dark matter (a term introduced by him: *matiere obscure*) must be less than or of the same order as ordinary matter. Jacobus Cornelius Kapteyn and Ernst Opik reached similar conclusions, while James Jeans and Jan Oort reached different conclusions: stellar motion required a mass greater than that observed.

Weighing a Cluster of Galaxies

A fundamental step in the path to the study of dark matter was taken by Fritz Zwicky in the 1930s, an astronomer with a Swiss father born in Varna, Bulgaria. Together with Baade, they coined the term supernova, and he hypothesised that neutron stars are formed in the collapse of these stars. Zwicky also hypothesised how cosmic rays could be produced by supernova explosions; furthermore, he proposed using objects more massive than stars to verify the gravitational lensing effect (see Chapter 4) predicted by general relativity.

In a work from 1933 and one from 1937, Zwicky studied the intrinsic velocities of the galaxies in the Coma Berenice cluster and used the velocities to trace the amount of mass that constituted the cluster. The structure of the cluster is given by the balance between the energy linked to the motion of the galaxies, called kinetic energy, and the gravitational force. If the kinetic energy is greater than that of gravity, the galaxies flee away from the cluster, while if the gravitational energy is greater than the kinetic energy, the galaxies fall towards the centre of the cluster. A stable cluster requires a balance between kinetic and gravitational energy. From the study of the motion of the galaxies in the cluster, Zwicky concluded that for the galaxies not to escape from the cluster, an amount of mass hundreds of times greater than that observed was necessary. He indicated this non-visible matter with the term *dunkle matter*, German for dark matter. The result was ignored for 40 years. A few years later, Smith, studying the Virgo cluster, confirmed Zwicky's result and assumed that the missing mass was made up of *internebular matter*. The following years were characterised by a long debate between supporters and detractors of the existence of dark matter. The debate was illuminated in the 1970s by the study of the components of clusters: galaxies.

The Legendary 1970s

The local mass distribution can be determined by using not only the vertical motions of the stars but also the circular motions of the stars on the galactic disc. Stellar velocity in terms of distance is referred to as the *rotation curve*. The study of the circular motions of stars in galaxies was begun by Babcock in 1939, who noticed that the stars in the outer part of Andromeda moved with an unexpectedly high speed, like the galaxies in the cluster studied by Zwicky. However, Babcock gave an interpretation that excluded the connection with dark matter. In 1959, Kahn and Woltjer, studying the approaching motion of our galaxy and that of Andromeda, concluded that the system contained more mass than that of all the stars.

Although the studies by Zwicky and Kahn and Woltjer clearly highlighted that clusters and galaxies contained more mass than visible mass, several doubts remained. The situation only became clearer in the 1970s.

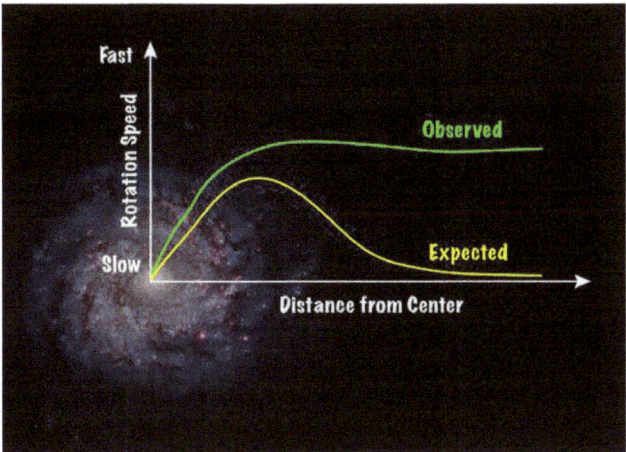

Figure 7.1. Typical rotation curve of a galaxy.

The pioneer of this change was Vera Rubin,[1] along with her collaborator, Kent Ford. The two began a study of Andromeda in 1970. Due to the rotational motion of the stars in the disc, the light coming from the stars approaching the observer undergoes a blueshift, and that from the receding stars shows a redshift. This is the well-known *Doppler effect*, and the change in the wavelength of light is proportional to the speed of the source. In this way, it was possible to determine the rotation speeds at different positions of the disc. As in the case of planetary orbits, Rubin and Ford expected an increase in rotation speed up to a maximum,[2] and then, moving away from the centre, a decrease in the speed. What they observed was completely unexpected: the rotation speeds of stars at large distances from the centre did not decrease but were similar to those of stars closer to the centre (Figure 7.1). The rotation curve had a flat structure.

---

[1] Vera Rubin became passionate about astronomy at an early age. She used to look at the stars and sky from her north-facing room in her home in Washington, DC. After finishing college, she was unable to continue her studies at Harvard because women were not accepted then, which was the case until 1975. She continued her studies at Cornell University, where she studied with famous physicists such as Richard Feynman and Hans Bethe. She obtained her doctorate from Georgetown University under the guidance of another famous physicist, George Gamow. She then secured a research position at the Carnegie Institution of Washington and began her observations there together with Kent Ford.

[2] Related to the central concentration of mass in the galaxy.

This was extremely strange because, by moving outwards, the stellar mass decreased, and with it, the rotation speed should have decreased. The flatness of the rotation curves of galaxies is strong evidence of the existence of dark matter. The galaxy must be contained within a huge region of non-visible matter. Several other researchers, such as Ken Freeman, reached similar conclusions.

In 1974, two groups led by Jaan Einasto and Jeremiah Ostriker stated in two important papers that the mass of galaxies was underestimated by a factor of 10. The result regarding excess matter was confirmed by Albert Bosma's 1978 study, which showed that using observations in the radio band, flat rotation curves were obtained at distances much greater than those studied in the optical band. The rotation curves are an important indirect proof of the existence of dark matter and also indicate how it is distributed within galaxies. Ordinary matter is more present in the internal part of a galaxy, and as we move away from the centre, it gives way to dark matter.

Taking an X-ray of Galaxies

Other evidence in favour of the existence of dark matter came from the revelation of emissions in the Coma and Perseus. In clusters, the gas is held by gravity. However, the gravity generated by the stars is not sufficient to maintain the gas. To prevent it from escaping, an enormous amount of invisible mass is needed: dark matter, in fact. The results confirmed previous estimates made by Zwicky. The results of the observations of the hot gas led to the conclusion that about 5% of the mass was made up of galaxies, 15% gas, and 80% non-visible matter. This matter is not interstellar material, which, despite not being visible optically, can be seen with infrared-sensitive telescopes, but matter which is not seen directly by any type of telescope.

Cosmic Mirages

As we saw in Chapter 4, the theory of general relativity predicts that space-time deformation also influences the propagation of light, deflecting the rays. Zwicky was the first to indicate, in 1937, that galaxy clusters could behave like gravitational lenses. More than 40 years had to pass before his intuition and the predictions of general relativity were verified

 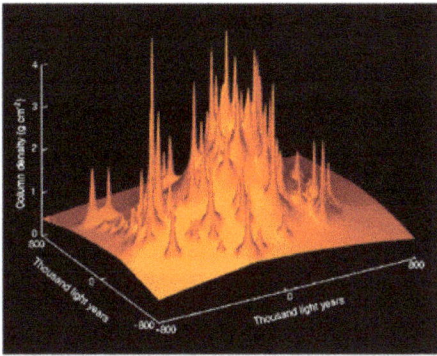

Figure 7.2.   Left: gravitational arcs produced by the image of a galaxy behind the cluster CL 0024+1654. Right: reconstruction of the mass of CL 0024+1654 in false colours. The plot represents dark matter per unit area, and the pinnacles represent galaxies.

Source: Greg Kochanski, Ian Dell'Antonio, and Tony Tyson (Bell Labs).

in 1979, when an object duplicated by the said effect was observed, the twin quasar SBS 0957+561. Many other confirmations of the gravitational lens effect have been described in Chapter 4.

The importance of the lens effect lies in the fact that, from the distortion of distant galaxies, it is possible to determine the total mass of the lens, i.e. the sum of the masses of the stellar material, the gas, and the dark matter. By measuring the stellar mass and that of the gas, the mass of dark matter can finally be determined. In 1990, Tyson and collaborators identified a coherent alignment of the ellipticities of blue galaxies behind the clusters Abell 1969 and CL 1409+52, and despite the weakness of the effect, using particular mathematical techniques and simulations, they reconstructed the mass distribution of the lens. Figure 7.2 shows, on the left, multiple images of a blue galaxy behind the CL 0024+1654 cluster and, on the right, the false-colour computer reconstruction of the mass distribution of CL 0024+1654. The peaks represent the contribution of galaxies to the mass. The image shows how most of the mass is found between galaxies and is on the order of 40 times greater than the visible mass of the system. Lensing is now routinely used to reconstruct the distribution of dark matter and can be considered a method for "seeing" dark matter.

Thus, there is a series of indirect evidence for the existence of dark matter based on the study of galaxies, clusters, and gravitational lenses.

There is various other evidence of its existence arising from the study of the way in which galaxies are distributed in space, from cosmic microwave background radiation, and from the collision of galaxy clusters. Describing them all would not bring us new knowledge and would only serve to confirm that there is matter that is not directly visible using telescopes.

Is Dark Matter Ordinary Matter?

We have seen a series of indirect evidence all converging on the claim that our Universe must contain dark matter. An open problem remains: understanding what dark matter is made of. The first idea that may come to mind is that it is ordinary matter that, for some reason, is not visible, for example because the radiation emitted is so weak that it cannot be detected. In 1956, Heeschen looked for neutral hydrogen (HI) emission in the Coma cluster and found it, but three years later, Muller showed that Heeschen's detection was spurious. In the mid-1960s, H. Rood tried to understand what dark matter was made of within galaxy clusters. He came to the conclusion that such matter must be found in intergalactic space. Since HI had not been observed, people began to think that it could be ionised hydrogen, but this hypothesis was discredited by observation. Having discarded these possibilities, attention was turned to massive collapsed objects with masses ranging from hundreds of millions up to thousands of billions of that of the Sun, a hypothesis discarded due to the absence of tidal distortion of the galaxies. Dwarf stars, planets, brown, red, and white dwarfs, black holes, and primordial black holes – these objects were given the nickname "MACHOs" (massive astrophysical compact halo objects) by Kim Giest. The idea that dark matter was made of MACHOs seemed like an excellent intuition, and for this reason, experiments were designed to verify the hypothesis. It was realised that the phenomenon of gravitational lensing could play an important role in the detection of MACHOs. Chang and Refsdal, in 1979, showed that stars, contrary to what Einstein had thought, could also act as gravitational lenses. In 1986, Paczinsky proposed using this effect to search for compact objects in our galaxy, and Nemiroff made calculations on the probability of microlensing events due to these objects. The micro-lensing effect is characterised by an increase in starlight depending on the mass of the object. For example, if we observe a star in the Magellanic Clouds and a MACHO passes through the line that connects the observer to the observed star, an increase in stellar brightness will be observed. The variation times range from a few hours to a year for MACHOs with masses between one ten millionth and one hundredth of a solar mass. Three collaborations, Experience pour la Recherche d'Objets Sombres (EROS), Optical Gravitational Lensing Experiment (OGLE), and Massive Compact Halo Objects (MACHO), began to search for MACHOs in the 1990s. The MACHO collaboration, after about six years and after the observation of the light curves of tens of millions of stars, revealed 14–17 microlensing events. The conclusions were that MACHOs must constitute between 8% and 50% of the mass of our galaxy's halo. After a year, the results led to a reduction in the percentage. MACHOs could constitute a maximum of 8% of the mass of the halo.

Primordial Black Holes

A final possibility of objects that could constitute baryonic dark matter are primordial black holes. In fact, black holes could also exist, not generated by the collapse of stars but generated in the first moments of the formation of the Universe due to the very high material density during those phases. Primordial black holes have quite small masses, $\approx 10^{12}$ kg, similar to those of a comet. In the first moments after the big bang, simple fluctuations in material density can give rise to regions so dense that they recollapse to form black holes. The idea that primordial black holes may constitute all or part of dark matter has become an active area of study following the detection of gravitational waves due to the coalescence of black holes by the LIGO/VIRGO collaboration. The black holes that produced the 2015 gravitational wave had unexpected masses for a black hole. One possibility is that they constitute a large family of black holes with a spatial distribution similar to that of dark matter, originating by accretion from the population of primordial black holes. The limits on baryonic matter given by primordial nucleosynthesis and the CMB would not apply to objects that formed before nucleosynthesis, at least for low-mass objects. The constraints on primordial black holes obtained with the EROS collaboration and from CMB anisotropies have been revisited, and some studies have shown that primordial black holes could constitute dark matter, while others have found opposite results. The matter is not very clear.

In conclusion, excluding primordial black holes, the theory of primordial nucleosynthesis places limits on the mass of baryonic matter in the Universe, which is 5%, as confirmed by the CMB study.

Furthermore, only about 1% condensed to form stars, planets, and other compact objects. Most baryonic matter is found in interstellar matter, in the hot gas inside galaxies and clusters. Consequently, these limits indicate that dark matter, which constitutes 26% of the matter in the Universe cannot be made entirely of baryons and, at the same time, that part of the baryonic matter is not visible.

Dark Matter in Particles

If dark matter is necessary to explain a whole series of phenomena, and if it is not composed of ordinary matter, i.e. baryonic matter, then what is it made of? Already in the 1970s, there was speculation that dark matter was made up of some sort of particle. Several reasons pointed in that direction. To begin with, baryonic matter did not explain the estimated amount of matter. Assuming that the Universe has Euclidean geometry, i.e. it is flat (see the following chapter), the estimates gave values for dark matter in the range between 20% and 30% of the mass content of the Universe, much greater than that expected from primordial nucleosynthesis, equal to approximately 5%.

The first particles that were linked to the missing mass were neutrinos. In 1973, Cowsik and McLelland, after having obtained an upper limit for the mass of neutrinos (1972), first proposed the idea that neutrinos with a mass of a few eV[3] could dominate the gravitational dynamics of galaxy clusters and the Universe and applied the result to the problem of missing mass in the Coma cluster. The most significant figure in neutrino cosmology is the Belarusian Yakov Zeldovich, an almost legendary figure of Soviet physics and astrophysics, who was highly versatile and prolific. He made notable contributions to fields such as nuclear physics, particle physics, relativity, astrophysics, cosmology, and materials physics, and he also played a crucial role in the development of the Soviet atomic bomb project.

---

[3] The eV, or electron volt, is a measure of the energy gained (or lost) by an electron, which moves in a vacuum between two points between which there is a potential difference of 1 V. The multiples of eV are: the KeV (1,000 eV), the MeV (1 million eV), the GeV (1 billion eV), and the TeV (1,000 billion eV).

Zeldovich's model of the formation of cosmic structures was based on neutrinos, which are particles with relativistic speeds, and for this reason, the model was called the *Hot Dark Matter* model (HDM, for short).

However, there is a non-trivial problem. In a universe dominated by neutrinos, the first structures to form are very large objects such as galaxy clusters, and galaxies are formed through *fragmentation*. It soon became clear that the neutrino-based model didn't work. In fact, galaxies form first in a universe and then clusters, contrary to the predictions of the HDM model. In 1982, when it was clear that the model of neutrino dark matter was not accurate, several physicists, including Jim Peebles, proposed that it was made up of other particles, such as axions, gravitinos, and photinos. Since these particles were non-relativistic, the model was called the *Cold Dark Matter model*, abbreviated to CDM. In it, structures form from small to large scales, producing structures in accordance with observations. In summary, as already mentioned, dark matter cannot be made of baryons, nor can it be made of neutrinos, because they are hot dark matter and have a very small mass. Although they are numerically very abundant,[4] they cannot constitute 26% of the critical density. Instead, the CDM model predicts perturbations in agreement with those observed in the CMB and does not have the problems of the HDM model.

We are again left with the question: What is dark matter made of? We think we know it must be made of particles, but in nature, there are different types of particles. What is the right particle?

Identikit of Dark Matter

Dark matter and its identification have many similarities to a detective story. It is very unlikely that when a crime is committed, investigators will be at the crime scene. Investigators, if they are lucky, may find clues. Each clue gives us information about the culprit. Once all the clues have been collected, a sort of identikit of the suspects can be drawn. By continuing this painstaking work, it is hoped that the culprit will eventually be caught. In short, using the words of Sherlock Holmes, "When the impossible has been eliminated, what remains, however improbable, must be the truth".

---

[4]To give you an idea, a surface of one square centimetre is crossed by 100 billion neutrinos in one second.

So, we need to compose an identikit of dark matter if we want to find it, and if we are talking about particles, we need to understand what type of particles they are, what range of masses they must have, and what type of interactions they feel.

We said at the beginning that baryons and neutrinos cannot be all the missing matter in the Universe. Following Sherlock Holmes' methods, we can try to see if any of the particles in the standard model are right for us. To examine whether a possible particle exists in the standard model, we must consider what the characteristics of dark matter particles must be:

- Dark matter is neutral; otherwise, it would interact much more than observations predict.
- It is subject to the gravitational force, while interaction through the other forces, if it exists, must be very weak; otherwise, it would have already been observed.
- Dark matter cannot be ordinary matter because limits from the primordial nucleosynthesis theory indicate that ordinary matter is less than 5%, much less than the expected amount of dark matter.
- Dark matter must be stable.
- Dark matter interacts very little with itself.
- Dark matter must be cold, i.e. of the CDM type.
- The particle constituting dark matter must be so abundant as to explain 26% of the mass expected from observations and theories.

We can now use the above criteria to see if there are particles in the standard model that are right for us.

A careful comparison between the identikit of dark matter and the particles of the standard model tells us that dark matter cannot be made up of usual matter.[5]

Despite this debacle, physicists have not given up on trying to find new dark matter candidates. The imagination of physicists has been so prolific that in recent decades, it has been seen that it is not difficult to propose candidates for dark matter; in fact, too many have been proposed. A class of particles proposed as dark matter candidates belongs to the

---

[5] There is an exception, namely the case of the axion, which is a dark matter candidate and is related to the standard model.

class of *supersymmetric particles*. Known particles can be divided into two groups: fermions and bosons. Fermions (e.g., quarks, electrons, and neutrinos) constitute the matter of which the world is made. Bosons (e.g., photons, W and Z particles, and Higgs boson) are associated with interactions, that is, they are the mediators of forces: electromagnetic, strong nuclear, and weak nuclear. Supersymmetry is a theory that corresponds to each particle with its supersymmetric particle, the *super-partner*. Each fermion corresponds to a boson and vice versa. The fermionic super-partner that has had the most success over time as a dark matter candidate is the neutralino.

The masses of neutralinos are in the range of 10 to 10,000 times the mass of the proton. This mass range is important to satisfy another characteristic required of dark matter particles: having an abundance in agreement with observations. In addition to neutralinos, there are other supersymmetric particles called weakly interacting massive particles (WIMPs).

WIMPs have been the trendiest particles for decades, but things have changed a bit in recent years. The reason is that the search for supersymmetry at the Large Hadron Collider (LHC) in Geneva did not give positive results. Other candidates for dark matter come from extra-dimensional theories. They assume that the number of dimensions of space is not three, as observed, but a larger number. We only observe three dimensions because the others are crumpled up into very small structures. If we want to enter one of them, we would not be small enough to be able to sneak in, but due to quantum mechanical effects, we would need an energy inversely proportional to the radius of the extra dimension. Since it is very small, the energy required would be enormous.

If we had two particles with enough energy to enter the extra dimension and we made them collide, they could move into the extra dimension after the collision. For an observer who does not see that dimension, he would see that the particles collide and remain stationary. Consequently, he would arrive at the incorrect conclusion that the principle of conservation of energy does not apply. That is, in the collision, the observer would think that energy was lost. So, we could get clues to the existence of extra dimensions in particle accelerators by studying the apparent loss of energy due to motion into the extra dimension itself. Experiments were carried out at CERN, but the desired effect was not observed. Each particle of the standard model that is in extra dimensions could generate heavier particles, not yet discovered; these particles constitute *towers* of particles of increasing mass.[6]

So, if we have an electron moving into extra dimensions, this will generate a Kaluza Klein (KK) tower of heavier electrons. The dark matter candidates constitute a veritable zoo. We focused more on some specimens belonging to the WIMP species because, in past years and even today, they have received considerable attention. Such supersymmetric particles have merits, as mentioned before, but unfortunately they have a big problem. After the non-observation of supersymmetry in the LHC experiments, doubts began to arise both on the existence of supersymmetry and, obviously, on its particles. Even if supersymmetry and its particles did not exist, there still remain a myriad of candidates for dark matter, such as *axions, sterile neutrinos, Wimpzillas, the dark photon, fuzzy dark matter,* and *Q-balls.* There are also scenarios such as self-interacting dark matter and asymmetric dark matter. In short, there is no shortage of candidates.

How to Find Dark Matter

In 2012, Katherine Freese and Christofer Savage published a paper on dark matter collisions with the human body, showing that billions of dark matter particles pass through us every second. A WIMP can hit one of the quarks that make us up with the exchange of a Higgs boson, and in a year, 10 WIMPs interact with our atoms. In other words, although the interactions of dark matter particles with ordinary matter are extremely weak, they are not zero. These particles obviously not only pass through us but everything around us.

So, to look for dark matter, we could build an experiment that attempts to verify whether WIMPs have interacted with nuclei. This method of searching for dark matter, proposed already in the 1980s, is called direct revelation. The probability that a dark matter particle interacts with a detector depends not only on its mass but also on its density. Dark matter is found everywhere in space. It increases going towards the centre of a galaxy. In Earth's surroundings, most mass is baryonic. In any case, a

---

[6]That is, the least massive particle, with mass m, is that of the standard model, and the others have mass $m_n = \sqrt{(m^2 + n^2/R^2)}$, where $n = 0$ is the particle of the standard model and the others with $n = 1, 2, \ldots$, are the other more massive particles. As you can see, the mass, m, depends on the size, R, of the extra dimensions.

coffee cup contains some particles of dark matter that are not stationary but move at high speeds, of the order of 300 km/s. In addition to the aforementioned method, there are two other methods of detecting dark matter: the first is based on the decay of dark matter particles into different possible particles, called indirect detection, and the other is based on the possibility that dark matter is generated and observed indirectly in particle accelerators.

Direct detection is based on the idea that a dark matter particle interacts with the nucleus of a certain material, producing a *recoil* of the nucleus which deposits energy in the material, which is then measured in different ways. A first possibility is the measurement of the vibrations of the crystal lattice. The second possibility is the detection of *ionisation*, i.e. electrons stripped from atoms. The third is *scintillation*, i.e. the generation of pulses of light emitted after the collision. There are dozens of direct detection experiments that measure one or more of the indicated effects.

Aside from the *direct revelation* method, a second possibility is *indirect revelation*. The foundations of this second method were laid in two 1978 articles by Gunn and collaborators and by Floyd Stecker. The two groups focused on heavy leptons (heavy neutrinos), showing how the annihilation of pairs of these particles could give rise to a flux of gamma rays. The study was extended to dark matter candidates of different types, such as WIMPs.

These particles annihilate, producing energies in the range 1–1,000 GeV (1–1,000 times the mass of the proton) and photons so energetic that they interact with the nuclei present in the atmosphere, giving rise to cascades of particles. To observe such gamma rays, there are two possibilities. The first is to use satellites that are above the atmosphere, and the second is to use ground-based telescopes that use the *Cherenkov effect*. When high-energy photons interact with the nuclei of the atmosphere, swarms of particles are generated, such as electrons and positrons, which move at a speed greater than that of light in the air, producing blue radiation contained in a cone: *Cherenkov radiation*, which takes its name from the first scientist who observed it. This light is collected by telescopes with large mirrors. Today, several of these telescopes are operational, including the High Energy Stereoscopic System (HESS) in Namibia, the Very Energetic Radiation Imaging Telescope Array System (VERITAs) in Arizona, the Major Atmospheric Gamma-ray Imaging Cherenkov Telescope (MAGIC) in the Canary Islands, Whipple in Arizona, and CANGAROO III in South Australia.

A satellite that has been of considerable importance to this research is the FERMI satellite. It was launched in 2008 and consists of a gamma-burst detector and the Large Area Telescope (LAT), which is sensitive to gamma radiation between 20 MeV and 300 GeV. In this second detector, the gamma photons are converted into electron–positron pairs, whose deposited energy is measured, and the direction of origin of the photons is reconstructed.

Another method to search for dark matter was proposed by Silk and Srednicki based on the study of antimatter (positrons and antiprotons) in cosmic rays. What is the connection between antimatter and dark matter? As seen, antimatter is contained in cosmic rays, with a ratio between positrons and electrons decreasing as the energy increases. In the annihilation of dark matter, however, the ratio between positrons and electrons increases with energy. So, to look for signs of the existence of dark matter, we could see if the ratio between positrons and electrons increases with energy in cosmic rays.

Instruments used for this research include PAMELA, FERMI, and AMS-02 installed in 2011 on the International Space Station, the Calorimetric Electron Telescope (CALET), and the Dark Matter Explorer (DAMPE), launched in 2015. These continue to search for dark matter but with results, unfortunately, so far not positive. Another method is based on neutrinos. The Sun can attract dark matter particles that remain trapped by the gravitational field, and when they collide with the solar nuclei, they lose energy and settle in the central part of the star. Neutrinos generated by the annihilation of WIMPs can leave the Sun and could be detected on Earth. This method was proposed in 1985 by Krauss and collaborators, and to put it into practice, several experiments were built, as follows: BDUNT[7] built inside the Russian Lake Baikal; ANTARES[8] in the south of France; and KM3NET under construction 80 km southeast of Capo Passero, off the southeastern coast of Sicily, at a depth of 3,500 m. Other experiments have been built in the ice of the South Pole: AMANDA and IceCube, built in a cubic kilometre of ice between depths of 1,500 and 2,500 m. The basic principle is common to all experiments.

When ultraenergetic neutrinos from the Sun give rise to electrons or muons in water or ice, moving faster than the speed of light, they produce

---

[7]Baikal Deep Underwater Neutrino Telescope.

[8]Astronomy with a Neutrino Telescope and Abyss Environmental Research project.

a Cherenkov flash of light. This radiation is detected by optical modules containing photomultipliers tied to strings lowered into water or ice.

Finally, another way to detect dark matter is to produce it in an accelerator like the LHC from very high-energy collisions between ordinary particles. Although dark matter interacts very weakly with ordinary matter, there is the possibility of generating dark matter particles from the collision of ordinary particles, for example pairs of protons. In the CMS and ATLAS experiments of the LHC, signs of the existence of new particles relating to different supersymmetric and extra-dimensional models were searched for. The results were all in agreement with the standard model. To be precise, in 2015, a signal was observed at the LHC that could be interpreted as the decay of a hypothetical 750 GeV particle into two photons. It was later shown that this was a statistical fluctuation.

Despite all the possible methods for detecting dark matter and the decades spent in its research, to date, there have been no positive results. Some experiments have led to uncertain results, especially those of indirect detection. For example, PAMELA had found signals compatible with dark matter, but it was shown that such signals could also be produced by pulsars.

Does Dark Matter Really Exist?

After decades of searches for dark matter without positive results, a spontaneous question is: Are there other possibilities beyond dark matter? Well, many scientists have dedicated themselves to providing a solution to this dilemma. We have seen that in the case of dark matter, the classic proofs of existence that we have discussed are all based on the idea that known physics and, in particular, Newtonian mechanics and Einstein's theory of relativity are correct. In reality, the latter has been verified on certain scales, for example in the solar system, but not on all scales. If it were not correct, it would be necessary to replace it with another theory of gravity. In recent decades, several theories of modified gravity have been proposed with the aim of explaining the Universe without the presence of dark matter. The basic idea is that, although experiments agree locally with general relativity, both in time and space, gravity could be different in the early Universe or on large scales. One of the first to propose the idea that we could get rid of dark matter by modifying gravity was Arrigo Finzi in 1963. His work remained forgotten, but two decades later, a similar idea was published by Mordeai Milgrom, the well-known

Modified Newtonian Dynamics (MOND), which changes Newton's laws at small accelerations of the order of 10^{-8} cm/s^2. In this way, it is possible to explain the rotation curves of galaxies without dark matter. MOND also explains several other things, but when you get to scales equal to or larger than those of clusters, problems begin. For it to work, it is necessary to reintroduce some dark matter in the form of sterile neutrinos. The situation is similar to that of a dog chasing its tail.

However, even the model of the Universe based on dark matter has problems, such as the difficulties in explaining the structure of small galaxies or the mass distribution in the collisions of some clusters. We mentioned that the distribution of mass in collisions between clusters is another evidence of the existence of dark matter. An example is the *Bullet Cluster*, 1E 0657-56. It is located in the constellation Carina and is the result of a collision at a very high speed (5,000 km/s), which occurred 100 million years ago between two clusters of galaxies, a larger one having a mass of 2×10^{15} M$_\odot$ and the other having a mass about 30 times smaller. The distribution of the masses that make up the galaxy cluster can be determined by studying the emissions in the X band and the gravitational lensing effect. Once the mass distribution has been reconstructed, we find ourselves faced with a unique spectacle: the baryonic mass that emits in the X-ray part of the electromagnetic spectrum is found in the central area of the cluster and is separated by two regions that contain most of the mass. In the collision, the gas, basically hydrogen, of the first cluster collides with the gas of the second, forming shock fronts at the centre of the cluster, which are visible by observing the X-ray emissions of the cluster. Using lensing, one can reconstruct the distribution of the dominant mass in the cluster, consisting of two mass concentrations localised symmetrically with respect to the centre. This mass has a completely different distribution from that of ordinary matter: it is dark matter that has not undergone interactions with ordinary matter. It continued its journey undisturbed, forming the two concentrations far from the centre, where the baryonic matter is found. Unfortunately, there is another cluster, Abell 520, in which dark matter is found close to the gas, contradicting the case of the Bullet Cluster and the basic theory of dark matter. In other words, the situation is complex: we have no certain proof of the existence of dark matter, and at the same time, we do not have a modified theory of gravity that could replace dark matter.

Chapter 8

What is Dark Energy?

Dark energy is incredibly strange, but actually it makes sense to me that it went unnoticed

— Adam Riess

A New Surprise from Cosmology

In Chapter 2, we saw that Friedman and Lemaitre, by solving the equations of general relativity, had discovered that the Universe was not static and that, in 1929, after the publication of Hubble's results, the idea that the Universe would expand. The famous Hubble law showed once again that the Earth and our galaxy were not in a privileged position in the Universe. This was a great discovery that had changed the paradigm of the static nature of the Universe of which everyone was certain. However, this was not the last big surprise that cosmology had in store for us. Sixty-nine years after Hubble's discovery, another certainty was demolished by the observation of a particular type of star, SNIa supernovae, as we will now see. If the Universe expands, the first question that can be asked is whether its rate of expansion has been and will always be the same or whether it changes over time. From an intuitive point of view, the answer that can be given is the following. Let us consider a certain region of the Universe containing galaxies. This region will tend to become larger due to expansion, but the force of gravity will tend to slow the expansion. Obviously, the more galaxies contained in the region, the less it will tend to expand. The pace of the expansion will therefore be linked to this battle between the force of gravity to keep the galaxies from moving away and the expansion. From an observational point of view, it is possible to determine the rate of expansion from the initial era to today. To this end, let's remember that when we observe a distant object, we observe it as it was when the light was emitted, not as it is today. By measuring the speed at which objects placed at different distances move away, we can determine the rate of expansion of the Universe at different times. To do this, we must determine the distance to the object and its speed of departure using independent methods. The constant of proportionality between speed and distance is the Hubble constant, which is precisely the rate of expansion of the Universe. The main problem is determining the distance to the object. How do we do it? We can use standard candles, objects of known brightness. We could think of using Cepheids, but they are visible just up to about 10 megaparsecs, i.e. about 15 times the distance between us and the closest galaxy to ours (Andromeda). This distance is not enough to establish what type of geometry our Universe has. To establish how the expansion occurs, we must go to distances of 3 billion 1,000 light years or greater. At such distances, the wavelengths are shifted by about 30% from the *cosmological redshift*, a time at which the Universe was about 10 billion years old. Furthermore, at distances greater than 3 billion light years, the expansion rate varies with time, and Hubble's law must be modified. The distance depends not only on Hubble's law but also on the density and pressure of the matter. A particularly important type of standard candle, which is right for us, is type Ia supernovae. They originate from the explosion of carbon–oxygen white dwarfs, devoid of hydrogen.

In the explosion of type Ia supernovae, the luminosity increases very rapidly and reaches an intrinsic luminosity equal to that of a few billion suns. Furthermore, the intrinsic luminosity is very similar in all explosions of this type due to the uniformity of the masses of the exploding dwarfs. A decay in brightness follows, and after a few hundred days, most of the SnIa disappear in the glow of the host galaxy. Therefore, the masses of the white dwarfs at the moment of the explosion are the same in all of them, and the physics of the explosion that gives rise to the supernova is similar in the different white dwarfs. Consequently, the energy released, the characteristics of the explosion, and those of the supernova phenomenon are very similar in different supernovae, which thus constitute standard candles. For the sake of precision, there are differences between the peak brightnesses of SnIa, but it is possible to correct these discrepancies using an empirical relationship found in 1993 by Mark Phillips between the peak brightness and the rate of decline of brightness with time: the faster the decrease in brightness, the weaker the supernova. In this way, SnIa can be made into perfect standard candles. They have the advantage over Cepheids of being much brighter and can therefore be observed from much greater distances. The second quantity we need is their receding speeds, which can be determined by studying the redshift of the light they emit. Once the distances and speeds of supernovae are known, we can evaluate the rate of expansion of the Universe at different times. However, there is a problem. SnIa are not frequent. A galaxy like ours produces a couple of them per century, and after they explode, their brightness is only visible for a few weeks. This means that it makes no sense to observe a distant galaxy and wait for a SnIa to explode; rather, we must observe many galaxies simultaneously in order to increase the probability of observing one or more supernovae. You have to observe thousands or millions of galaxies simultaneously. In the 1990s, two projects with this aim were created: the *High-z Supernova Search Team* led by Brian P. Schmidt and the *Supernova Cosmology Project* led by Saul Perlmutter. Perlmutter's project was born from a redirection of a project on mass extinctions produced by astronomical causes. In 1984, palaeontologists David Raup and Jack Sepkoski argued in a paper that there was a 26-million-year periodicity in mass extinctions. Two groups of astronomers, Whitmire and Jackson and Davis, Hut, and Muller published hypotheses similar to those of the aforementioned palaeontologists in the same year. According to them, the mass extinctions would have originated from a dim star called Nemesis, named after a goddess from Greek mythology. Such a star would be a red or brown dwarf, which would perturb the objects found in an immense sphere around the solar system, called the *Oort cloud*, made up of an enormous number of comets. As a consequence, it would increase the number of comets moving towards the solar system with an increase in the probability of impact with the Earth. Perlmutter's supervisor was one of the astronomers mentioned, Muller. Perlmutter had to create an automatic faint star search project. The project led to nothing, and in order not to throw away the work, it was suggested to convert the project into an automatic search for supernovae. The other team's story is less peculiar. Observing approximately one million galaxies per night, the two teams discovered several SnIa located at different distances. By comparing these distances with the relative receding speeds, the two projects were able to calculate the change in the rate of expansion of the Universe from remote times until today. An interesting thing is that the goal of the two teams was to measure the slowdown of expansion according to the dominant ideas at the time. The first challenge they faced in determining the rate of expansion of the Universe was to build a catalogue of supernovae at distances varying up to a few billion light years. Phillips' supernovae were too close, so with painstaking work, the two groups had to build the necessary catalogue. It took them five years to do this, resulting in supernovae up to 10 billion light years across. Now that they had the required supernovae, they could know how much the Universe had expanded since a given time in the past by considering their distances and the redshift z of light from distant galaxies, which provides, among other things, their speeds. To establish the rate at which the expansion occurred, we need to know the distances to the galaxies together with their redshifts. If the rate at which the Universe expands decreases over time, this means that expansion was faster in the past, and therefore the Universe took less time to reach its current state. As a result, the Universe is younger than a universe with accelerated expansion. Light therefore has to travel a shorter distance to reach us. Since a standard candle is brighter the closer it is to us, if the expansion rate decreases, we will have supernovae that are brighter than those in a universe in which expansion occurs at a constant rate. Now, what astronomers observed was exactly the opposite: the supernovae were less luminous. In other words, for a given redshift z, the supernovae were further away than would have been the case if the Universe had expanded at a constant rate.

So, the Universe was accelerating. Another fundamental point is that a universe made of only matter cannot produce accelerated expansion. In fact, gravity tends to slow down the expansion of the Universe; therefore, there must be a new component that gives rise to gravitational repulsion. For this purpose, a fluid that has negative pressure is needed. This new component of the Universe is referred to as *dark energy*. The two groups came to the conclusion that a new component of the Universe was needed, trying to explain the data they had collected. In the analysis of the two groups, three different models were considered. Two models consisted of matter alone, baryonics and dark matter, one with a density parameter $\Omega_m = 0.3$, and one with a density parameter $\Omega_m = 1$. Astronomers saw that the supernovae were dimmer than slow-expansion models predicted. To explain the data, they introduced a third model which, in addition to the material component with $\Omega_m = 0.3$, took into account the presence of a cosmological constant with $\Omega_\Lambda = 0.7$, interpretable as a fluid at negative pressure. In such a cosmology, the Universe has accelerated and expansion was slower in the past; therefore, the Universe took longer to reach

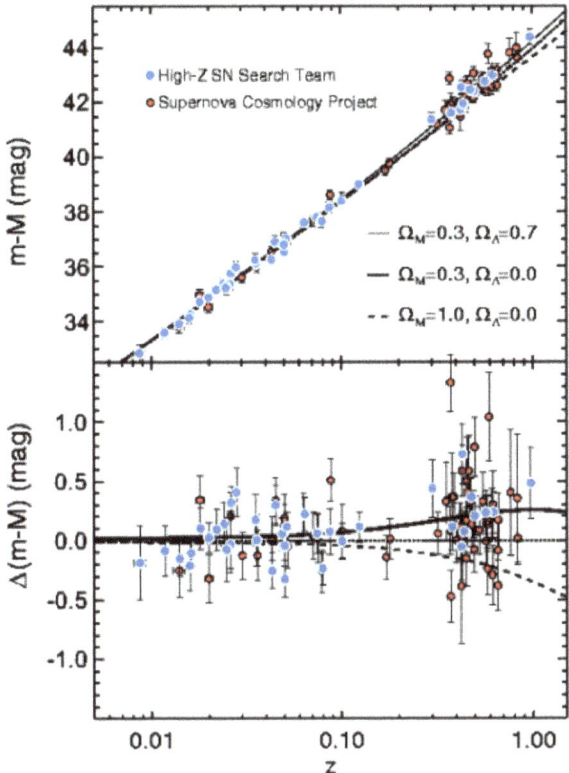

Figure 8.1. Left: Results from the High-Z SN Search Team and the Supernova Cosmology Project. The simplest form of this entity that accelerates the Universe, dark energy, is the cosmological constant Λ.

its current state. Consequently, the Universe is older than in other cosmologies. A second important point is that light must travel a greater distance to reach us, and consequently, supernovae, like other sources, are less luminous than in a universe with slow expansion. The data from the work of the two groups are illustrated in Figure 8.1, which represents the distance, linked to brightness, as a function of redshift.

At first glance, you don't see big differences between the cosmologies with a slowing-down universe (the curve with a broken line and the one with dots), the first two lines from the bottom, and the accelerated one, the continuous curve at the top. Looking carefully, you can see that the SnIa are distributed more on the top line, corresponding to an accelerating universe.

Cosmological constant as Dark Energy

To obtain the accelerated expansion of the Universe, we need a component different from ordinary matter, which we have indicated with dark energy, and we know that $\Omega_\Lambda = 0.69$. The term Λ is used for historical reasons. Einstein, in 1917, in an article that marked the birth of modern cosmology, "Cosmological Considerations on the Theory of General Relativity", introduced this constant to make the Universe static. Even with Λ, the Universe was static for only a moment and then either collapsed or expanded. That is, with Λ, the Universe was unstable. After Hubble's discovery of the expansion of the Universe, Einstein said that using the cosmological constant was the biggest mistake of his life. Despite his declaration, the cosmological constant has had changing fortunes since Einstein's time. In some periods, it has been forgotten, while in others, it has been revived, and today it is considered a panacea to explain the accelerated expansion of the Universe. To tell the truth, already in Einstein's time, Willem de Sitter had studied a Universe without matter but with the cosmological constant and had noticed that it gave rise to a repulsive gravity. Furthermore, this gravity was uniform in time and constant in space, and as we will now see, another oddity of the cosmological constant gives rise to a negative pressure. So, the interesting thing is that the deprecated cosmological constant is able to explain the accelerated expansion of the Universe. How does it do it? The equations of general relativity are made up of a left-hand part which describes the geometry of space-time and a right-hand part linked to the mass and energy contained in it.[1] To explain the accelerated expansion, Λ is added to the right-hand side of the equation and can be interpreted as a constant contribution to the energy density of the Universe. It is an energy that uniformly fills the entire Universe, dark energy.

Λ produces gravitational repulsion, and consequently there is an accelerated expansion of the Universe. To understand why this happens, we must remember that in general relativity, gravitational forces are not only produced by matter and energy but also by the pressure exerted by the previous ones. If the pressure is negative, repulsion is produced. Suppose we have a container with an ideal gas, and we compress it. To reduce its volume and compress it, we have to make an effort, and the gas accumulates energy. Suppose we have the same container full of dark energy.

---

[1] The equation, for visual reference only, is $R_{\mu\nu} - 1/2 g_{\mu\nu} R + \Lambda g_{\mu\nu} = 8\,\pi G/c^4\,T_{\mu\nu}$.

Logically, this situation is not real since dark energy extends everywhere, and it is not possible to harness it in a container; however, we suppose it is possible to see what would happen. Since the density of dark energy is, by definition, constant, the energy in the container is proportional to the volume it occupies. By lowering the piston, as the volume decreases, the energy also decreases. Dark energy has the typical behaviour of a system with negative pressure, or negative density. In a system with negative density, gravity behaves in the opposite way to how it usually does: it behaves as antigravity and produces a gravitational repulsion. Consequently, dark energy can explain the accelerated expansion. To explain the expansion in terms of what is observed in our flat Universe, it is necessary to have $\Omega_\Lambda = 0.69$. The antigravitational and repulsive force of dark energy modifies the evolution of the Universe. As can be seen in Figure 2.1, there is a decelerated expansion up to around 5 billion years, and then the expansion accelerates (dotted line). In general relativity, the cosmological constant is a simple term added to the field equations. One may wonder what the physical meaning of this quantity is. One answer comes from quantum mechanics and quantum field physics. The cosmological constant Λ is today interpreted as the energy of the quantum vacuum. But there is a problem with this interpretation. If you do the calculations with quantum field theory, you find that the discrepancy between the observed values of the cosmological constant and the calculated values of the vacuum energy amounts to a factor of 10^{120}, a huge number, which constitutes the *problem of the cosmological constant*.

The cosmological constant has another problem, called the *coincidence problem*. Until about 5 billion years ago, the Universe was not accelerated because there was much less dark energy than today. Since then, it has increased so much that it has a value similar to that of dark matter, and in the future, if dark energy is truly the cosmological constant, there will be a further increase. The *coincidence problem* can be stated as follows: Why are the densities of dark energy and dark matter comparable today? This could either be a coincidence, or there could be some fundamental reason why this is the case. The history of Λ is long and complex, as described in a 2017[2] Raifeartaigh article. It was not exhumed in 1998 because, after it was pulled out of the hat by the magician Einstein, it never wanted to go back in again.

---

[2] https://arxiv.org/ftp/arxiv/papers/1711/1711.06890.pdf.

The Quintessence

So, in summary, a non-zero cosmological constant justifies the accelerated expansion of the Universe but suffers from several problems, such as the aforementioned problem of the cosmological constant. Given this and other problems, alternatives to dark energy have been proposed. An alternative that enjoys particular interest is the *quintessence*, or *fifth element*, which, in the Aristotelian worldview, constituted the spheres and celestial bodies, ranging from the lunar sky to the fixed stars. The history of the ether began with the ancient Greeks, Plato, and Aristotle, who gave it its name, ether, and a systematic treatment. Unlike the four elements of the sublunary world, earth, water, air, and fire, ether was the essence of the celestial world and indeed the fifth element. In the Middle Ages, the term ether was transposed into quintessence, which was the constituent of the philosopher's stone, which, in mediaeval beliefs, was a substance with phenomenal properties, among which we remember the ability to confer immortality. The ether also had success in the 19th century, when it was postulated by supporters of the wave theory of light to explain its propagation in a vacuum. Its non-existence was highlighted by Michelson and Morley's experiment in 1887, which opened the way to Einstein's theory of special relativity. Today, the term quintessence comes from the fact that the Universe is made up of four components: dark matter, baryons, photons, and neutrinos; therefore, dark energy constitutes the fifth element. Thus, quintessence is a form of energy hypothesised to explain dark energy. Unlike the cosmological constant, which does not change with time, quintessence changes over time and perhaps even across space. The quintessence is usually represented by a perfect fluid with negative pressure, P, and positive density, ρ, which follows the law $P = w\rho$, called the *equation of state*, and w is called *the state parameter*. The state parameter is generally negative, so one has a negative pressure. When w equals -1 ($w = -1$), we are dealing with the cosmological constant. There are models with $w < -1$: the *kinetic quintessence* (*k-essence*), with a non-typical form of the kinetic energy, and the *phantom energy* model, in which the kinetic energy is negative and the expansion is faster than that of the cosmological constant. A universe dominated by phantom energy would lead to the *big rip* (which we will talk about in the following chapter), an end similar to that which Jack the Ripper reserved for his victims. Quintessence has also been proposed by several physicists as the fifth force. This force would modify the evolution of the cosmos. Until now, we have only known four forces, and to explain what happens around us or in the solar system, we do not need a fifth force. Consequently, there must be a *"chameleon-like" mechanism* that shows the effects of this force only in certain conditions and hides them in others. For example, it must hide them in our laboratories and in the solar system, where the observations are well described by Newton's equations. This hypothetical phenomenon is called *screening*, and there are different types (Chameleon, Vainshtein, and so on). However, all these theories have no experimental confirmation, and the old cosmological constant continues to be the most probable reason why the Universe expands in an accelerated manner.

Dark Energy: Beyond Supernovae

Besides supernovae, there are other ways to show that our Universe requires an additive component to explain observations. As we saw at the beginning of this chapter, the geometry of space-time is determined by the amount of matter and energy contained in it, as described by the density parameter Ω. There is a value for the density parameter for each material component of the Universe. For example, the one relating to dark matter is $\Omega_{dm} = \rho_{dm}/\rho_c$. In the calculation that leads to the value of Ω, the various forms of matter in the Universe must be taken into account. From the study of the microwave background radiation, it is possible to trace the Ω values for each component. According to data from the 2015 PLANCK satellite, the contributions due to baryonic matter, Ω_{bar}, are equal to 0.0486 (4.86%), those of dark matter, Ω_{dm}, are equal to 0.2589 (25.89%), and those, negligible, of radiation, Ω_{rad}, 0.00005 (0.005%) and neutrinos, Ω_v, 0.004 (0.4%). Summing the indicated contributions, we find that $\Omega_{matter} = \rho/\rho_c = \Omega_{dm} + \Omega_{bar} + \Omega_{rad} + \Omega_v \approx 0.31$. Again, from the study of microwave background radiation, we know that our Universe is flat, which means that the value of Ω_{Total} must be approximately equal to 1 ($\Omega_{Total} \approx 1$). This implies that a matter-energy component equal to 0.69 (69%) is missing. As is known, this is dark energy. The sum of all contributions, including dark energy, gives $\Omega = \rho/\rho_c = \Omega_{dm} + \Omega_{bar} + \Omega_{rad} + \Omega_v + \Omega_\Lambda = 1.0023 \pm 0.0054$. This result tells us that the Universe is flat and infinite and that dark energy contributes 69% of the matter-energy of the Universe. Gaining a better understanding of dark energy has become one of the most important problems in current cosmology. To understand the nature of dark energy, it is necessary to use a variety of methods to measure acceleration. A first method independent of supernovae is that of *barion acoustic oscillations* (BAOs). What is it about? The Universe, until 380,000 years after the big bang, was made up of a plasma of baryonic matter and photons. The regions of higher than average density attracted matter to themselves, and the photons produced a pressure that opposed gravity. This tug of war between gravity and pressure created oscillations, similar to sound waves. The regions of higher than average density were made up of baryons, photons, and dark matter, and the pressure gave rise to spherical sound waves of photons and baryons that moved from the central region of greater density, made up of dark matter, outwards at a speed of 170,000 km/s. Acoustic waves travelled in the primordial plasma until they cooled and formed neutral atoms 380,000 years after the big bang. The photons were free to move and moved away at the speed of light, while the baryons remained "frozen" in a spherical region. So, the spherical sound waves, before vanishing, left an imprint of their existence on the matter of the Universe. The baryons located in the said spherical regions and the dark matter at the centre of the regions constituted the regions of inhomogeneity that attracted the matter, forming galaxies. Sound waves travelled from the big bang to recombination for 380,000 years. The distance that the sound travels in this period is indicated by the term *sound horizon*. Also, taking into account the expansion of the Universe, the spherical regions at recombination had a size of 450,000 light years. The current size is 490 million light years. As a consequence, we expect to see a greater number of pairs of galaxies separated by 490 million light years compared to a random distribution. Like the SnIa, the BAOs provide standard candles, thanks to which it is possible to determine the distance. More precisely, they provide a standard rule of 490 million light years for length scales in cosmology. It can be measured by studying the large-scale structures of the Universe through surveys, such as the Baryon Oscillation Spectroscopic Survey (BOSS). BAOs can improve our knowledge of acceleration by comparing observations of the sound horizon today, using the way galaxies are distributed, to those of the sound horizon at the time of recombination, using background radiation. Therefore, BAOs provide a sort of ruler of known and standard dimensions with which we can better understand the nature of the acceleration, completely independent of supernovae.

Is Accelerated Expansion Produced by Dark Energy?

One question that can be asked is whether there are other possibilities for producing accelerated expansion beyond dark energy. After the discovery of the accelerated expansion of the Universe, in addition to the cosmological constant and the quintessence, a large number of *theories of modified gravity* were proposed, which indicates that, especially in the case of dark energy, there are unclear ideas. One of these theories is, for example, the *DGP modified gravity theory* (from the initials of the authors: Gia Dvali, Gregory Gabadadze, and Massimo Porrati). Differently from the theory of general relativity, in this theory, gravity is produced by particles called gravitons, which decay. Due to their decay, gravitational attraction decreases, and the expansion of the Universe accelerates. However, there is no evidence that this theory, like other theories of modified gravity, is correct. In addition to theories of modified gravity, another possibility is that, on a large scale, the Universe is neither homogeneous nor isotropic and that the Earth is located in a huge empty region, and the rate of expansion would vary based on location, which could be interpreted as varying over time. Expressed in a slightly more technical way, it is usually assumed that the Universe is homogeneous and isotropic; however, in reality, on scales smaller than a few hundred million light years, this is not true. Therefore, the inhomogeneity could influence the way in which the light spreads. Studies on a particular model of the Universe (Lemaitre–Tolman–Bondi) have shown that inhomogeneity mimics acceleration, i.e. inhomogeneity is interpreted by us as accelerated expansion. This action of small-scale structures on the large-scale behaviour of the Universe is called *feedback*. Other doubts about the fact that the Universe expands in an accelerated manner were recently raised by Jacques Colin and collaborators, who showed that the acceleration is not the same in all directions and that the acceleration deduced from supernovae would not be real but rather appears so due to the fact that we are non-Copernican observers. An open discussion on this point would prove Colin wrong.

However, most scientists agree with the idea that, to explain our Universe, we need dark matter and dark energy. Until direct evidence of the existence of dark matter and a greater understanding of dark energy are found, we must remain open to all possibilities. One of the hopes is the EUCLID mission, recently launched, which could clarify these issues.

Chapter 9

What Will Be the Destiny
of the Universe?

This is how the world ends. Not with a Bang but in a whimper.

— Thomas Stearns Elliot

The questions of how the universe began and how it will end are probably as old as human civilisation. Obviously, our approach to solving these questions has changed considerably with the evolution of populations. The questions about the origin and end of everything are closely linked to those about the origin and end of the individual, that is, life and death. The answers to these questions have been linked to mysticism and religions for millennia. As societies have evolved, more philosophical and rational arguments have been sought. There are myriad eschatologies, doctrines aimed at investigating the destiny of the individual, of the human race, and of the universe. Just to give a few examples, the mythologies of the Vikings and other Nordic peoples were completed with the *Fimbulvetr*, a very cold three-year winter followed by the *twilight of the gods*, the final struggle between the forces of order and those of chaos. This would have led to the end of the world, which would have then been reborn and repopulated. In Hindu cosmology, Brahma is the creator of the world, and a day of Brahma, or Kalpa, lasting 4.32 billion years is a measure of the cosmic cycles of evolution and involution of the universe. At the end of each day of Brahma comes a night of Brahma, during which the world is partially destroyed by fire, water, and wind. After 100 years of Brahma, Brahma dies, and the universe is completely destroyed and will not exist for the next 100 years of Brahma. Then, Brahma will be reborn and with him the world in a continuous cycle. The eschatologies, like the Viking one, are similar to today's cyclical theories of the universe. Unlike the various and picturesque eschatologies of the peoples who have lived on Earth, today cosmology allows us to answer in a scientific, even if not certain and unambiguous, way the question relating to the end of the Universe. In reality, there are several possibilities, and these fundamentally depend on the geometry of the Universe and its future evolution, which depend on the relative proportions of matter and dark energy, the nature of which is not yet known. In any case, the main current cosmological models predict either a universe that had a beginning and will have an end or a universe that is cyclical, that is, one with an end, but at that end, there is a new start.

In the past, there was a theory, called the *steady state theory*, according to which the Universe was infinitely extended in space and time: it had no beginning and would not have an end. This theory was definitively refuted by the observation that the universe was different in the past and, above all, by the discovery, in 1965, by Arno Penzias and Robert Wilson, of the cosmic microwave background radiation, a sort of sea of residual microwaves that pervades the universe and reaches the Earth from across the sky, a true imprint of the big bang. It was predicted in 1948 by Alpher, Herman, and Gamow. Recalling Chapter 2, we know that Friedman had shown that the Universe, depending on the matter content, can either expand and then collapse again in a big crunch or it can expand forever. From cosmology, and in particular from the study of microwave background radiation, we know that we live in a flat Universe, with a density equal to 5 hydrogen atoms per cubic metre and, therefore, that the Universe tends to expand forever. The presence of dark energy, with its current behaviour, produces an acceleration of the Universe. In conclusion, by knowing the density of the Universe and the behaviour of dark energy, we can know what the future of our Universe will be.

In particular, if the behaviour of dark energy will always continue to be the same and if its density will remain constant as described by the cosmological constant, what we expect is that the Universe will expand forever and that all the galaxies will move away until they are no longer visible. The future that awaits our Universe, in this case, is *heat death*, which we will talk about shortly. This is the most likely end for our Universe. If dark energy did not behave as described by the cosmological constant and was characterised by higher than a certain value, more

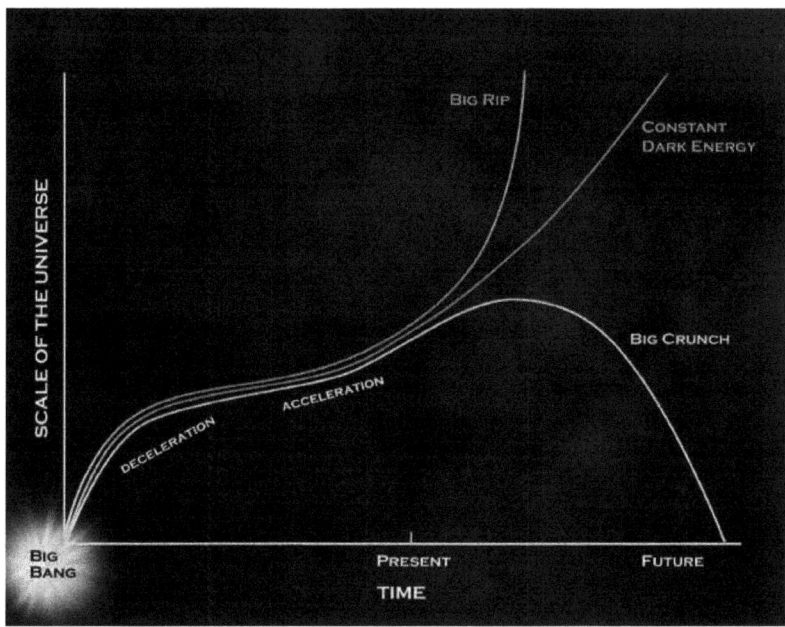

Figure 9.1. Possible fates of the Universe. The accelerated expansion continues as it does today if dark energy is constant, or a very accelerated expansion occurs due to the growth of dark energy, leading to the big rip. Finally, we have the case of contraction and recollapse (big crunch).

Source: Adapted from NASA/CXC/M. Weiss.

precisely if the parameter w was less than -1, the Universe would not reach the stage of thermal death; instead, before reaching it, the matter in it would be torn to pieces. This is the *big rip* scenario. If energy weakened or changed the way it behaves, the deficit of matter and energy would cause the collapse of space, as we have indicated with the *big crunch*. Figure 9.1 summarises these possible fates for our Universe.

Apart from these possibilities, there is another possibility, linked to ancient philosophical ideas: the *cyclical model*, that is, a Universe that is born, dies, and is reborn. For example, in *Thus Spoke Zarathustra*, Nietzsche speaks of the eternal return, the eternal repetition of all events in the world and all realities. Nietzsche recovers the cyclical vision of time of Pythagorean and Platonic reminiscence, as opposed to the linear Christian one. The idea of a cyclic Universe was introduced by Friedman himself when he obtained his solutions to general relativity. He proposed that the Universe originates, expands to a maximum, and then collapses again in a big crunch, and this will be followed by a new big bang and the repetition of the process *ad infinitum*. Today, there are several other proposals for oscillating universes. In the following sections, we deal with the possible fates of our Universe, starting with the most probable one: *heat death* or *big freeze*.

Heat Death (Big Freeze)

We mentioned that our future will most likely end in a heat death, that is, a state of the Universe in which entropy reaches its maximum value and the Universe reaches thermodynamic equilibrium. The first ideas on the thermal death of the Universe are linked to William Thomson, who proposed them in 1851. This ending of the Universe is linked to the second principle of thermodynamics, according to which an isolated system's entropy, or disorder, increases irreversibly. If the universe lives long enough, it will reach a state of uniformly distributed energy, making the existence of energy processes, including life, impossible. In the case of heat death, the Universe would expand, stars in galaxies would extinguish, black holes would evaporate, and protons, if some theories are correct, would disintegrate. The universe would become an immense, cold void populated by photons and with energy distributed uniformly. As mentioned, today this is the most accepted idea on the end of the universe, even if there are conflicting points of view on this model, on the applicability of entropy and thermodynamics to the entire universe, and on making predictions based on poor knowledge of the entropy of gravitational fields and the role of quantum effects. Nonetheless, let's assume that this is how the Universe will end. What will happen? We could build a timescale for the end of the Universe and life on Earth. Due to the increase in solar brightness, in about six hundred million years, the level of carbon dioxide will drop below the level necessary for the production of C3 photosynthesis used by most plants. Almost all plants will die, except those that use carbon-4 (C4) for photosynthesis. However, even these plants, along with the animals depending on them, will not survive long. In about a billion years, solar brightness will increase by about 10%, and the Earth will be at the mercy of an uncontrolled greenhouse effect. The oceans will slowly evaporate. Unless humans learn to make interplanetary travel and find another place to live, they will be doomed. Extremophile bacteria or tardigrade aquatic invertebrates no larger than half a millimetre in size could perhaps continue to survive on Earth. Events unfavourable to life will follow one after another regularly. In about three billion years, the Large Magellanic Cloud, one of our small satellite galaxies, will collide with ours, and a billion years later, there will be a collision with the galaxy closest to us, Andromeda. Assuming the solar system survives these events, in about five billion years, the Sun will run out of hydrogen and transform into a red giant, expanding about a hundred times. If the oceans have not already completely evaporated due to the greenhouse effect already mentioned, they will, and the rocks will melt. In our galaxy, star formation will end in about 5×10^{10} years. After a few tens of millions of years from the formation of the last stars, those with masses greater than 8 solar masses will explode in the form of supernovae, and after 10^{12}–10^{14} years, the stars will extinguish. The disappearance of these stars, which tend towards blue, will lead to the galaxy taking on a yellowish colour; furthermore, heavy elements will no longer be formed. Since the brightness of a star tends to increase during its existence, for a few million years, even after the cessation of star formation, the brightness will remain constant. Then, all the stars will form fossils such as white dwarfs, neutron stars, and black holes, and the galaxy will be made up of these objects, along with brown dwarfs and planets. The temperatures of all objects will decrease as they move around in their orbits. The stars will begin to leave their orbits as two objects orbiting one another emit gravitational waves. It is estimated that due to gravitational radiation, most of the orbits of the stars in the galaxy will decay, and they will be swallowed up by the central black hole. The galaxy will be filled with stellar fossils: white dwarfs, neutron stars, and black holes. These objects will also be swallowed up by the central black hole over time, apart from a few lucky objects, which, through interactions with others, will be able to acquire enough energy to escape from the central regions of the galaxy. In 10^{19} years, the stars will separate from the galaxies, and in 10^{20} years, the orbits will decay. The Earth–Sun system will probably survive a little longer, up to 10^{23} years. Not only will planetary systems and stars already be dead, but even galaxies will begin to evaporate. Not even ordinary matter is eternal. Leptons (e.g. electrons) are stable particles, whereas neutrons decay in about 15 minutes, but it is not known whether protons decay or not. The future of the Universe will take different turns depending on whether protons are stable or not. Some theories predict the decay of the proton in 10^{33} years. To verify these predictions, various experiments were carried out. The most recent experiment studying this problem is the *SuperKamiokande experiment*, which used 50,000 tons of ultra-pure water containing 1,034 protons and more than 11,000 photomultipliers that reveal the blue light produced by the *Cherenkov effect*, already mentioned. According to the experiment, the halflife[1] of the protons must be greater than 10^{33} years. The SuperKamiokande results do not confirm that the proton does not decay but only give a lower limit to its halflife. So, maybe protons will decay in more than 10^{34} years, or maybe they won't decay. If they decay, the decay products, such as positrons, will annihilate with the electrons. The material of the Universe would be transformed into iron, and this into neutrons. Even protons joining electrons would form neutrons. In an unimaginably long number of years, all matter would be swallowed up by black holes. These black holes won't live forever either but will slowly lose mass due to *Hawking radiation* (see Chapter 5) and disappear in a sort of explosive emission of photons. A black hole of stellar origin with 10 solar masses evaporates in 10^{70} years, supermassive ones at the centres of galaxies with a mass of one million solar masses do so in 10^{85} years, and one with a mass of 10^{11} solar masses evaporates in 10^{100} years. So, all the energy and mass of the universe will be absorbed by black holes, which will eventually evaporate, emitting photons. This last conclusion is valid if Hawking radiation, which has never been observed, really exists. Black holes will absorb the mass and energy of the universe, and when they evaporate, only photons will remain, as well as perhaps electrons that have not disintegrated upon encountering positrons generated by the decay of protons. So, in the classical heat death (big freeze), the universe will be reduced to a cold and empty land in which photons, and perhaps electrons, wander. To this basic scheme, other scenarios have been added that could occur after the *big freeze* or before it occurs, changing the final fate of the universe. One of the eventualities related to the big freeze is the possibility of the end of time. Ultimately, time had an act of birth with the big bang and could therefore end, leaving everything that the universe will consist of in a frozen state, as in a still image. The imagination of physicists is very prolific, and there are therefore other possibilities that avoid heat death or that allow the Universe to go beyond heat death. If the vacuum that the Universe is made of was a false vacuum, i.e. a state with energy greater than the minimum energy corresponding to the true vacuum, after an enormous time ($10^{2,500}$ years), the false vacuum could decay

---

[1]Halflife is the time it takes for the number of protons, or other radioactive elements, to be reduced by half.

into a true vacuum, which could even have a big bang. This scenario is the same as those of universes dominated by a false vacuum, which can lead to the sudden disappearance of the Universe in a big slurp, as we see in Appendix 1.

In summary, the classical thermal death of the Universe predicts that we will arrive at a cold universe containing only photons, but some ideas would presuppose that a new universe, even cyclical, or a multiverse could originate from it.

An Empty Universe

We now wonder how living beings living in the Universe will see this. For this reason, we assume that dark energy continues to have the properties it has now. In this case, we will see galaxies continue to move away from our galaxy with increasing speeds until they exceed the speed of light. As strange as this may seem, this is not contrary to the predictions of relativity, which tell us that the maximum possible speed is that of light. In our case, it is space that is dilating, as space is created between galaxies at a speed greater than that of light, and this is not in contradiction with relativity. When the relative speed of galaxies exceeds that of light, we will no longer be able to see them. In fact, the light that leaves them will move at a speed lower than the speed at which the galaxies move away, which will therefore be invisible and will leave our horizon. They will not suddenly disappear from our horizon, but their light will be subject to the cosmological redshift: we will first see them become redder and redder, and then their light will shift towards infrared, microwaves, radio waves, and so on. This effect is due to the fact that wavelengths lengthen due to the expansion of the Universe.

The time for this to happen is about 10 times the current lifetime of the Universe. Not all galaxies will disappear. The galaxies of the local group, which would be our galactic neighbourhood made up of about 70 galaxies with a radius of about 1.5 Mpc, will continue to be visible because they are gravitationally linked to us. In about 100 billion years, our sun will have already gone through the red giant phase and become a white dwarf that will have cooled into a black dwarf. Our galaxy, after colliding with Andromeda and other galaxies, will be different. There will be stars and perhaps even civilisations on some planets related to them. These populations will not see the Universe as we see it today. Apart from the galaxies in their surroundings, they will see nothing and will conclude, like our 19th-century astronomers, that the Universe is static and consists only of our galaxy and its neighbours. They won't even be able to understand that the Universe is expanding. The three fundamental proofs of the big bang will no longer be available. The first, the expansion of the Universe observed by Hubble, will no longer be visible as only the galaxies of the local group will be able to be observed, which, following various collisions, will form a single enormous galaxy. We would no longer have evidence of the existence of dark energy, despite the fact that it far dominates the density of the other components of the Universe, because we will not observe either expansion or acceleration. The second proof of the big bang, the existence of the cosmological background radiation, will no longer be available. In fact, in a universe a few hundred times the current one, given that the temperature decreases with the size of the universe, the temperature of the background radiation will be a factor 100 times smaller and hundreds of millions of times less intense. The enormous difficulty of observing this radiation will be complicated by the presence of interstellar plasma. The third proof, the abundance of light elements, would however remain. At nucleosynthesis, the amount of hydrogen and helium was about 75% and 25%, not much different from today. As time grows, the quantities of helium and metals[2] will be much greater than today, and there will be no need to assume a big bang to agree with observations.

This description of the future of the Universe, which is nothing other than the heat death scenario, is based on the hypothesis that dark energy does not change its behaviour. If this were not the case, we would have completely different scenarios, as we see in the following sections. If dark energy will maintain its nature, our possible descendants would have to invent a way to describe an apparently static Universe, as Einstein already mistakenly did. The future described in this chapter is the possible future, a future that could change, as that of a human being does depending on one's choices. As Niels Bohr said, *Making predictions is a very difficult thing, especially if they concern the future.*

Big Rip or Big Crunch?

As already mentioned, the end of the Universe depends fundamentally on the unknown nature of dark energy. If it is not a cosmological constant but

---

[2] In astronomy, all elements heavier than hydrogen and helium are referred to as metals.

a dynamic quantity, i.e. it can change over time, different endings are possible. If the vacuum energy will grow over time, another scenario that has a similar ending to that of the big freeze is the *big rip*, proposed in 1993 by Robert R. Caldwell, Marc Kamionkowski, and Nevin N. Weinberg. This would happen in the case of a particular form of dark energy called *ghost energy*, characterised by a state parameter, *w*, i.e. the ratio between pressure and density, $w = P/\rho$, being lower than -1. In this case, a much more accelerated expansion would be produced than in the case of a cosmological constant, as shown in Figure 9.1.

The expansion of space would be so rapid, due to the repulsion produced by dark energy, that no force in nature, not even the strong interaction, could counter it. As a consequence, all the structures that make up the Universe, starting with galaxies, stars, and planets, up to molecules, atoms, hadrons, and so on, would be broken down into elementary particles: photons, leptons, and protons (if the latter were not subject to decay). This event could happen, according to some theorists, in 20 billion years. Subsequent events would be similar to what was described in the case of the big freeze, with the end of time and space. However, current studies on dark energy lead us to think that it probably does not have the characteristics of ghost energy. For these reasons, it is more likely that the universe is dominated by dark energy, such as a cosmological constant, and the universe would end up in a big freeze. Saying it with Thomas Stern Elliot, *This is the way the world ends. Not with a bang but with a whimper*. If dark energy did not grow aggressively and its nature changed completely, and instead of being repulsive, it became attractive (*w* > 0), there would be the opposite possibility to that of the big rip, that is, the universe would expand up to a maximum and then collapse again into the big crunch (see Figure 9.2). One possibility is that the big crunch is followed by a big bang, generating a cyclic universe, which in any case in standard cosmology has problems with entropy, as we discuss in the following section.

The Great Bounce and the Cyclic Universe

As we said previously, in the pursuit of solving the equations of general relativity, Friedman, in 1922, found three solutions, one of which described the universe as cyclical, i.e. constituted by a succession of big bangs and big crunches, similar to the situation of a ball that moves away

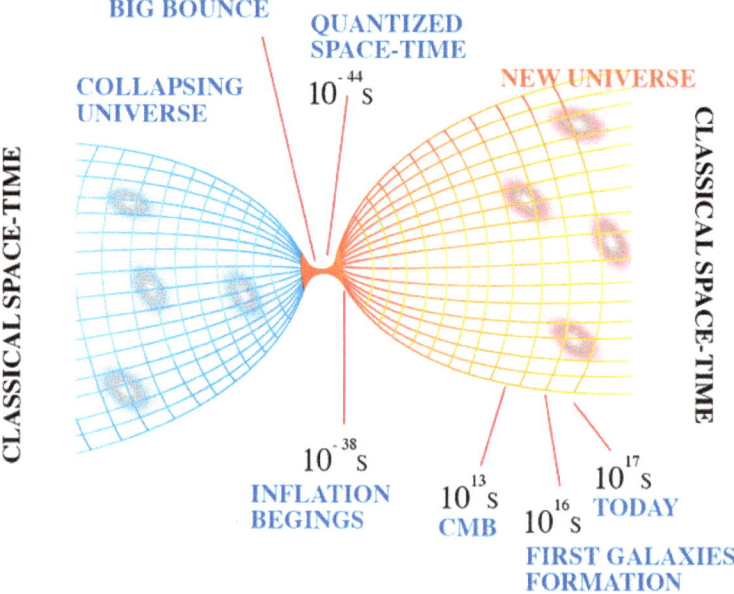

Figure 9.2.   Big bounce: the Universe, contracting to the left, reaches a minimum size, and then a new expansion occurs.

Source: Adapted from "Astronomica Mens", https://astronomicamens.wordpress.com/2016/08/16/big-bounce-unipotesi-alternativa-sullorigine-delluniverso/.

from the ground up to a maximum height, followed by a fall towards the ground and a bounce (*big bounce*).

This was also one of the models designed by Richard Tolman. Each bounce would have generated a singularity not describable by general relativity. One of the problems of the model is to explain what happens to entropy, i.e. the degree of disorder, in the various phases. The initial entropy of the Universe, at the first big bang, was extremely low and grew as its size increased. In the recollapse phase, entropy does not reverse its course, as galaxies do, but as Tolman pointed out in 1931, it will continue to grow. Consequently, for each cycle, there will be an increase in entropy, with a consequent increase in the maximum dimensions and the time between subsequent cycles. If we want a cyclic universe, it is necessary to find a way to decrease entropy. In 1999, Paul Steinhardt and Neil Turok believed they had found a possible solution based on superstring theory (see The Multiverse of String Theory). In this theory, particles are represented by oscillating strings. Different particles correspond to oscillations with different energies. An extension of the theory predicts that membrane-like structures of different sizes also exist. These are called branes (see The Multiverse of String Theory). The Universe could be a brane, floating like a sheet in hyperspace with a higher number of dimensions. The space-time fabric is made up of the three-dimensional surface of the Universe. Steinhardt and Turok imagined two brane universes, each consisting of nine spatial dimensions, but only three of the nine dimensions are visible and the other six are compactified. The universes were immersed in a 10-dimensional hyperspace. The two branes are bound together by a closed string, which carries gravity, and separated by an extra dimension that could contract periodically (see Figure 9.3).

Given that the gravitational force and the particles that mediate the interaction, the gravitons, propagate between two branes, and given the contraction of the extra dimension, the two brane universes attract each other, giving rise to a collision called *big splat*. This event predates the big bang and is a fundamental phase in the formation of a cyclic universe. This universe was called the *ecpyrotic universe*.[3] According to the two authors, the entropy problem was solved because only the extra dimension contracted and not the branes, which continued to expand. The entropy created in the branes spread and was never concentrated. Dark matter and dark energy were also explained by this model. Dark matter would be the manifestation of the gravitational force of another brane and the ordinary matter it contains. The collision between the branes would produce enormous energy and heat, which, generating a gigantic explosion, would give rise to a new universe in accelerated expansion. The latter would have originated from the residual dark energy of the collision. After the *big splat*, the branes move apart and come closer together again due to gravity and the cyclic contraction of the extra dimension, giving rise to a new cycle. Differently from the classical model of the big bang and inflation, the ecpyrotic model completely eliminates the singularity. Dark energy is a sort of extra gravity that, among other things, keeps the branes aligned and stabilises them. The branes are not rigid, but they are like sheets in the wind, so the collision occurs at different points and times. At the points

---

[3]This term can be translated as "transformation into fire" or "coming out of the fire", a term which, in Stoic philosophy, indicated the moment in which the world was cyclically created and destroyed.

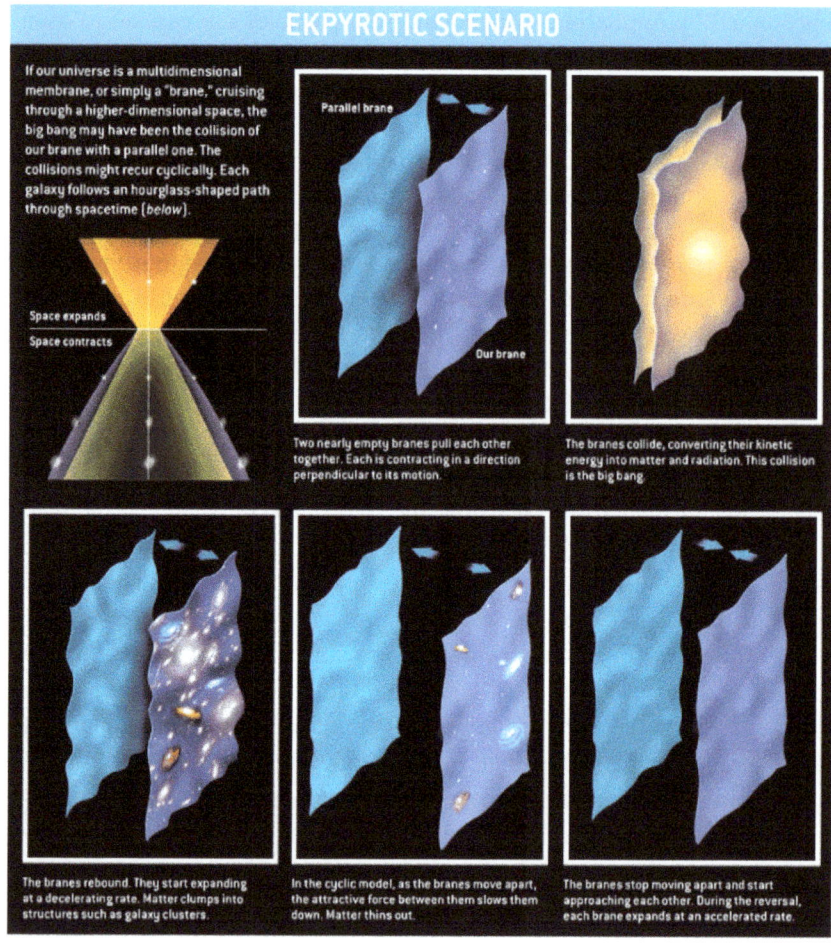

Figure 9.3.   The ecpyrotic Universe.

Source: Reproduced from https://westernparadigm.wordpress.com/2010/07/14/the-ekpyrotic-scenario/.

where it occurs first, the matter and energy produced by the collision thin out due to the expansion. At the points where the collision occurs afterwards, the inhomogeneities that will eventually give rise to the galaxies originate. The big bounce has also been brought back into the spotlight by superstring theory and the so-called loop quantum gravity theory. According to the cosmology of superstring theory, also called *pre-big bang cosmology*, the big bang was not the origin of everything but just an instant in which a state of very high density performed a sort of rebound into a state of rapidly decreasing density. In superstring theory, there is a fundamental length that can be thought of as the size of a point in space. This dimension is equal to 10^{-34} m, which ultimately represents the smallest radius of space. Superstring cosmology describes a flat and nearly empty Universe in its early stages, with similar characteristics in the future. Our Universe emerged from an eternally existing pre-Universe, when the density reached its maximum value in a collapse phase (the big crunch), which from our point of view is a big bang. This was followed by a phase of exponential expansion and the accelerating Universe we observe today. So, in superstring theory, there is no beginning.

The theory of *loop quantum gravity* aims to build a single theory, starting with general relativity and quantum mechanics. In it, space-time is made up of rings 10^{-35} m in size, which can contain a finite amount of energy. Therefore, as in superstring theory, there is a minimum dimension below which one cannot go. In quantum loop theory, space is discrete. Taking quantum mechanics into account, a contracting universe cannot be squeezed beyond a certain limit, just as an electron can only approach the nucleus up to a certain distance. It is as if quantum mechanics introduced a repulsive force that produces a great bounce. The said theory, similarly to the superstring theory, does not admit the singularity of the big bang. The predicted universe is a variant of the *oscillating universe*. The universe expands to a maximum and then collapses again. We go from a pre-universe of eternal duration, a bounce to time zero, and an infinite universe. The universe described by the theory has no beginning or end.

To conclude, another type of cyclic universe was proposed by Roger Penrose, the so-called *conformal cyclic cosmology*. According to Penrose, in a universe that has reached thermal death, the microscopic could have influences on the macroscopic, causing the dead universe to give rise to a new big bang. The entropy problem is solved by assuming that information, and therefore also entropy, is lost in the final evaporation of the enormous black holes that dominate the universe after 10^{100} years. Furthermore, the model predicts that the collision between the black holes should produce gravitational waves visible as concentric rings on the CMB. According to Penrose and Gurzadyan, such rings are visible in the CMB map of the WMAP satellite.

Another possibility for the end of the Universe is related to the decay of the vacuum, the *Big Slurp*, described in Appendix 1.

Chapter 10

Are There Other Universes?

In an infinite multiverse, there is no such thing as fiction.

— Scott Adsit

One question we might ask ourselves is whether there is something beyond the Universe we observe, or whether there even exist other universes of which our is only an example. Science fiction has produced an innumerable number of films on the subject: *Source Code, Everything Everywhere All at Once*, and many others. As early as in 1848, Edgar Allan Poe wrote a poem about the existence of a large number of universes, and before him, Giordano Bruno had spoken of infinite worlds. The multiverse is a convenient narrative device, but in recent decades, science has also become interested in it. The concept of the multiverse has gained increasing prominence as modern scientific theories have predicted the existence of other universes while attempting to explain our universe, such as cosmic inflation (which we will talk about), or extend certain points of view within theories like quantum mechanics. Another motivation that led to the introduction of the multiverse is the problem of the "fine tuning" of constants. The physical constants present in our Universe (intensity of forces, mass of particles, etc.) seem to be fixed in such a way that life appears in our universe. Tiny changes in some of these constants would have resulted in generating a completely different universe in which we would not be present. There are therefore different types of theories of the multiverse, arising from new interpretations of quantum mechanics, ideas about the primordial Universe, or theories that attempt to unify gravity and other interactions, such as superstring theory. In the following, we discuss these possible multiverses.

The Many-Worlds Interpretation of Quantum Mechanics and the Multiverse

The idea of a multiverse first appeared in quantum mechanics and was proposed by Hugh Everett III in 1957. Everett proposed a new interpretation of quantum mechanics called the *many-worlds interpretation*, which differed from the *Copenhagen interpretation of quantum mechanics*. He attempted to eliminate the problem of wave function collapse and reduce the role of the observer, central to the Copenhagen interpretation.

Before describing Everett's interpretation, we must remember the differences between the mechanistic, Newtonian vision of the world and the revolutionary perspective that quantum mechanics brought to physics.

Pierre Simon de Laplace, in *Exposition of the System of the World* (1796), developed (parallelly to Kant) the hypothesis of the origin of the solar system from a primitive nebula. He gave the text as a gift to Napoleon, who, after reading it, asked him why God was not mentioned in the entire text. Laplace replied, "Sire, I had no need for this hypothesis".[1] Laplace's conclusions were based on determinism, the idea that the phenomena that occur in the world are linked together by precise cause–effect relationships. In this way of seeing the world, which reigned until the 20th century, probability was excluded and everything could be described rigorously using mathematical and physical laws. The deterministic Universe was strictly governed by causality. Returning to Laplace, he was convinced that, using Newton's laws and knowing the initial conditions of the Universe, it was possible to predict every past and future state of the same. This idea of the world was completely shattered by quantum mechanics. It replaced determinism with indeterminacy and brought probability and "strangeness" into physics. *Heisenberg's uncertainty principle* states that quantities such as position and momentum (the product of mass and velocity) are subject to uncertainty. The greater the precision with which we know one of them, the less precision with which we know the other. Indeterminacy does not depend on our limitations in conducting

---

[1] It seems that Napoleon, upon meeting Lagrange, told him what had happened, to which Lagrange replied, "It's true. Yet it is a beautiful hypothesis that explains many things".

experiments; it is intrinsic to nature. Observation modifies reality. While in the macroscopic world there is an objective reality, in the microscopic world things happen in a completely different way. The attempt to define an objective reality is destined to fail since the action of the observer changes it into a subjective reality. Moreover, the observer is not a passive spectator; our presence changes reality. This highlights that reality is shaped by the observer. So, the world that quantum mechanics shows us is completely different from the one Laplace had in mind. Delving into quantum rules implies abandoning the intuitive schemes on which our vision of the world is based, built day after day through our interactions with the macroscopic world. The other aspect that makes quantum mechanics different from classical mechanics is that it is a theory entirely based on probability. After the first formulation of quantum mechanics by Heisenberg, which proved difficult for physicists due to the use of tables of numbers called matrices, Erwin Schrödinger tried to find a quantum theory following Newton's basic idea of formulating an authentic mechanics that could explain both the motion of bodies and the causes. Using the known results, he attempted to obtain a formula analogous to Newton's second law in the quantum world. Schrödinger hypothesised the existence of a function, called a *wave function*, capable of containing all the information relating to the system. This function is the solution of the *Schrödinger equation*, which, like Newton's second law, is a differential equation, a form of equation often used to express physical laws. Initially, it was not clear what the meaning of the wave function was. It was given by Born, who showed that the squared modulus of the wave function gave probabilistic information on the position of the object under study. The certainty of classical mechanics was replaced with probability. Some physicists were uncomfortable with the introduction of probability and the interpretation of the theory itself. If we can only measure probabilities, if the uncertainty principle places limits on measurements, if wave–particle duality tells us that a system can behave like a wave or a particle, we wondered where the objectivity of physics had gone. Bohr and Heisenberg, the fathers of the so-called *Copenhagen interpretation of quantum mechanics*, argued that the wave function represents everything we can know about a system, and observation and measurements were the only real things. For example, it would not make sense to ask where the particle was and what it was doing before the measurement. When we observe the particle and carry out the measurement, it chooses (almost like a thinking being) its position. The wave function is said to collapse to a precise value.

Before the collapse, the particle existed in a superposition of all possible states. For many, the most critical point of the Copenhagen interpretation was, as mentioned, the problem of measurement and the micro-macro problem. The latter is based on the fact that objects in the quantum world have different behaviour from macroscopic ones.

Macroscopic reality is independent of measurement, while microscopic reality depends on the act of measurement. Since there is no physical or logical threshold that separates the micro from the macro, this problem must be explained. That is, at what spatial scale does quantum behaviour end to give way to classical behaviour? Erwin Schrödinger, in an attempt to clarify this problem and ridicule the paradoxes of quantum mechanics, proposed the *paradox of Schrödinger's cat*, shown in Figure 10.1. Basically, you have a cat in a closed container together with a Geiger counter containing a radioactive substance in such a small quantity that you do not know whether one of its atoms will disintegrate or not within an interval, say one hour. If an atom were to disintegrate, the counter detects it and activates a hammer, which breaks a vial containing cyanide. You wait an hour, and if you don't open the box, not knowing if an atom has disintegrated, you can't know if the cat is alive or dead. It would therefore result in a state suspended between the living and the dead. As Schrödinger wrote, "The function of the entire system leads to the affirmation that in it the living cat and the dead cat are not pure states, but mixed with equal weight".

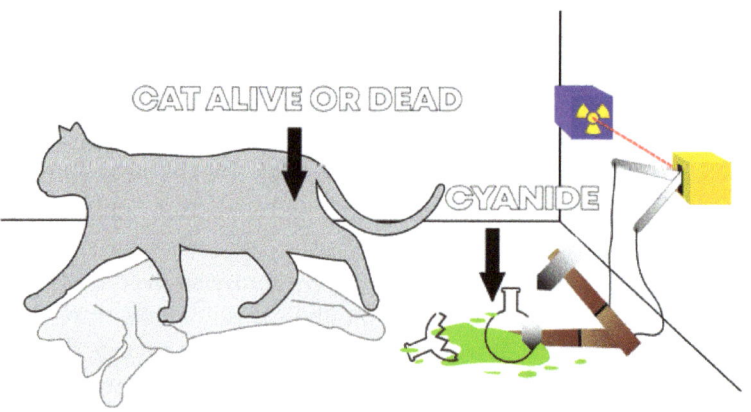

Figure 10.1.   Schrödinger's cat experiment.

Source: Gianni Comini, Rules of the game in quantum theory.

In summary, when the cat is enclosed in the box, it is in a superposition of two states: a live cat and a dead cat. When the box is opened, i.e. when the measurement is carried out, the wave function will collapse and reality will materialise before the eyes of the observer: the cat will be either alive or dead. At this point, a problem arises: what happened to the unobserved alternative possibility? Did it vanish the moment you opened the box? According to Niels Bohr, this state vanished. At a macroscopic level, no superposition of states is observed, as the Schrödinger's cat paradox clearly shows: a living being cannot be both alive and dead at the same time. In 1927, Bohr and Heisenberg established that the act of measurement causes the system to pass from the mixed state, or the superposition state, to a single state, i.e. the observed one. That is, carrying out a measurement on a quantum system destroys the superposition, making it "classical". The mechanism that produces this phenomenon, introduced by von Neumann, is called, as already mentioned, the collapse *of the wave function*.

This interpretation of quantum reality, as mentioned earlier, is known as the *Copenhagen interpretation*. This interpretation states, in other words, that reality is created by measurement. The role of the observer is fundamental in creating reality, but this goes against common sense, namely that the Universe must exist even when we are not looking at it. To solve these problems, in 1957, Hugh Everett proposed the *many-worlds interpretation of quantum mechanics*, in which the situation is not as para- doxical as in the Copenhagen interpretation. In Everett's many-worlds interpretation, the observer and the measured system constitute a single state called "world". Upon measurement, the global state is divided into as many worlds as the results of the measurement. In simpler terms, if you carry out an experiment that has two possible outcomes, the Universe will split in two. In Schrödinger's cat experiment, the act of measurement causes the Universe to split into two: in one, the cat is alive, and in the other, it is dead. Every event is a branch point for the Universe. The many- worlds interpretation has had detractors since its origin, starting with Bohr. The splitting of our world in two every time a decision is made or some- thing happens is difficult to accept. The idea that there is a Universe in which Caesar never existed and another in which he lived thousands of years ago and all the other possible and strange combinations does not intuitively testify in favour of the many-worlds interpretation. However, in the 1970s, when additional dimensions began to appear in physicists' cal- culations, they thought that these could host parallel universes located in a dimension that we cannot perceive. Everett's many worlds could have been those parallel universes. Today, this large number of universes is referred to as the *multiverse*. Returning to the measurement and the phenomenon of the collapse of the various possibilities, before the measurement, into a single reality, after the measurement, things were clarified after the mid-1990s with the theory of *decoherence*. The mechanism described by Bohr and Von Newmann raises questions such as: Assuming that there is a boundary between the quantum world and the classical world, where should it be drawn? Why should such a border exist? What is meant by observer? In his formulations, Bohr had certainly not clarified what an observer was. For example, in the case of Schrödinger's cat, is the observer the cat or the human who opens the box? The answers to these questions came with the theory of decoherence. According to this theory, quantum mechanics must be applied to "isolated" microscopic or macroscopic systems. If a quantum system is not isolated from the outside during a measurement, it is correlated with the environment, which must also be considered using the rules of quantum mechanics. If a system is in a coherent superposition of states, i.e. the quantum states cannot be written as a mixture of other states, the correlation with the environment leads to a loss of coherence between the different parts of the wave function relating to the superimposed states. We spontaneously pass from the *coherent* to the *decoherent state*. In the latter, the system is no longer in a superposition of states. The difference between microscopic and macroscopic systems is that while the former can be isolated from the environment, this is not so simple for macroscopic systems. These interact with the environment, and it is impossible to observe the superpositions of states. Furthermore, by observer, we must not mean a sentient being but only an interaction between particles. In fact, in 1996, Serge Haroche and collaborators showed, using rubidium atoms and microwaves, that decoherence is produced by the interaction of the atoms with the environment.

A Multitude of Multiverses

The multiverse of the many-worlds interpretation of quantum mechanics is the first and one of the possible types of multiverses. There are other forms of the multiverse. Max Tegmark carried out a sort of classification of multiverses. There would be four different types of multiverses:

- **Type I**: Universes similar to each other, unobservable, with similar physical laws and similar to our universe. Our flat Universe is probably infinite and could be a type I universe, as we will see in the following section.

- **Type II**: Similar to type I but with different physical law constants and spatial dimensions. This type of multiverse can arise in one of the forms of inflation theory. We will talk about this in the section "The Multiverse of Chaotic Inflation".
- **Type III**: With the same characteristics as type II and related to the many-worlds representation of quantum mechanics, discussed in the previous section.
- **Type IV**: Different forms of the laws of physics. Every mathematical structure has a correspondence in the physical world. The multiverse that arises from string theory is of this type. We will discuss this in the section "The Multiverse of String Theory".

Beyond this classification, Brian Green, in his book *The Hidden Reality: Parallel Universes and the Profound Laws of the Cosmos*, mentions nine different types of multiverse. To recall some of them: one of the types would be the *landscape multiverse*, based on Calabi–Yau spaces, about which we talk in the following; the *quantum multiverse* linked to the many-worlds interpretation of quantum mechanics by Hugh Everett III (mentioned a little earlier); the *brane multiverse*, based on *M* theory, where each brane would constitute a universe; the simulated multiverse that exists on very powerful computers, which would simulate entire universes – an idea similar to that of the film *The Matrix*; the *holographic multiverse*, based on the *holographic principle*, which asserts that our three-dimensional Universe would be a sort of projection of a two-dimensional reality, like a hologram. Extending the idea to the multiverse, we obtain the holographic multiverse. The dynamics of the inflationary multiverse would be contained within its outline. The list goes on.

As you can see, there is certainly no shortage of proposed multiverses. The problem is that it is much easier to imagine their existence, but it is certainly much more complicated to be able to demonstrate that they exist.

Infinite Universe and the Multiverse

We know that our Universe is flat and could even be infinite. Consequently, only a part of the entire Universe can be observed, the so-called *observable Universe*. We cannot observe further because the expansion of the

Universe occurs at a superluminal speed. If space-time has no end, it will have to repeat itself eventually.

Moving in an infinite Universe, sooner or later we should find a clone of the Earth and its inhabitants. According to Max Tegmark, the closest copy would be found at $10^{10^{29}}$ m. Moving a little further, $10^{10^{120}}$ m, you could even find a clone of our Universe. The probability of running into a clone of ourselves, although theoretically possible, is unlikely given that in a Universe based on the Big Bang, the separation between us and our clone is greater than the size of the horizon, which means that not only will we not meet our clone, but we won't even be able to communicate with them. In Tegmark's classification, this would be a *type I universe*.

The Multiverse of Chaotic Inflation

As we will see in this section, a variant of inflation theory, *chaotic inflation*, predicts the continuous birth of universes from the origin of our own Universe. These parallel universes, as already mentioned, have been classified as *type II universes*.

Before the 1980s, cosmology suffered from certain problems, the most burning among which were the following:

- **The problem of the horizon**: Today, we observe a homogeneous and isotropic universe, as seen in the background radiation. How did this condition originate from regions that were very distant from each other, not causally connected, and had never had contact?
- **The problem of flatness**: Today, the Universe has a density that differs from the critical one by less than 1%, and in the early Universe, this discrepancy must have been less than one part in 10^{61}. How can this fine-tuning situation be explained?
- **The problem of topological defects and the fate of space-time aberrations in the Universe**: An example is magnetic monopoles, hypothetical particles with only one pole which should be present in the Universe but are not.

These problems were solved in one fell swoop by a theory, called *inflation theory*, initially proposed in 1979 by Alexei Starobinski and a few years later by Alan Guth. In short, the central idea of inflation is that after the origin of the Universe around 10^{-35} s, the Universe underwent an exponential expansion that took it from microscopic to macroscopic dimensions, increasing its radius by a factor of 10^{25}–10^{30}. There is not just one model of inflation but several; however, they are all based on the hypothesis that inflation is driven by an unknown scalar field called *inflaton*, precisely because it produces inflation, i.e. the enormous expansion of the Universe. The concept of field is known to us from the concepts of electric and magnetic fields. A *scalar field* differs from these because, while electric and magnetic fields are entities that associate more than one number with each point in space, a scalar field associates only one number with each point. An example of a scalar field is the temperature field in a room: in one position, we have one temperature; in another, we have another temperature.

A complete description of inflation requires the use of complex concepts from field theory and particle physics. Here, we give an intuitive description of how inflation occurred. A more complete description is developed in Appendix 1. Inflation occurred at the *Grand Unification phase transition*, i.e. the separation of the strong force from the electroweak force when the Universe had an age of 10^{-36} seconds. Before seeing what happened, let's make a comparison with water when it is supercooled. We know that when the temperature of water drops to 0°C, a phase transition begins, and it turns into ice. However, if we slowly lower its temperature, we can avoid the transformation into ice and bring the water to temperatures of about 10 degrees below 0°C. There is therefore a state with a lower energy density, that of ice, available for the molecules of supercooled water. This state is more convenient from an energy point of view. However, supercooled water continues to remain in the liquid state, having a higher energy density than ice. So, our liquid system is in a condition of high symmetry, given that water is highly symmetrical – it appears the same from whatever direction you look at it. If the system is suddenly disturbed, the phase transition begins and the water becomes ice, releasing energy, as already mentioned, the so-called *latent energy*. The system passes to a lower symmetry since the ice is made up of crystals and has privileged directions within it. Something similar happened to the Universe.

According to Guth, at the end of the Grand Unification era, at times of the order of 10^{-36} s and temperatures of the order of 10^{28} K, in which the electroweak and strong forces were unified, the Universe entered a state of supercooling, with a temperature much lower than 10^{28} K, and remained trapped in a particular state called a *false vacuum*. We talk about a false vacuum when a system has an energy level higher than the minimum, the vacuum, or a *true vacuum*. In other words, the Universe was not in the state of minimum energy, i.e. the true vacuum, but had energies greater than those of the true vacuum. The situation is similar to that of water, which is in a state of higher energy density than it would have had if it had transformed into ice.

So, the Universe was in its state of false vacuum, with high symmetry and high energy density, despite the fact that a more favourable state existed, that of true vacuum, with lower energy density and lower symmetry. Inflation, or the exponential expansion of the Universe, started when quantum fluctuations pushed a small region of space into the true vacuum. The bubble made up of the true vacuum had zero pressure, but the false vacuum surrounding it had negative pressure. The increased pressure in the true vacuum bubble caused it to expand very rapidly, producing inflation. According to Guth, the size of the Universe could be 10^{23} times larger than that of the observable Universe, but there are very different estimates, both larger and smaller. At the end of the inflation, fields and particles were generated, and the region in which the inflation occurred began to expand at today's rate. In the final phase of inflation, the virtual particles that appeared and disappeared in the Planck era transformed into real particles due to the release of vacuum energy. The empty Universe became filled with particles and, therefore, mass. In Guth's terms, the Universe would be likened to *a free lunch*, or, in other words, it would be generated from the void. This is reminiscent of the *ex nihilo* creation supported by some religions. The remaining problem is to understand whether or not the vacuum coincides with nothingness.

Guth's inflation theory has problems determining the end of inflation. For this and other reasons, Guth's model was replaced by other models (proposed by Andrei Linde, Paul Steinhardt, and others), in which inflation is slower and the Universe is not initially trapped in a false vacuum. In these new models, once inflation begins, there will always be an exponentially expanding region that will give rise to another Universe. This is one of the variants of inflation, *eternal inflation*, in which, in different regions of the universe, inflation continues forever. Andrei Linde proposed a theory based on eternal inflation called *chaotic inflation* or *bubble theory*, which can be described as follows. The Universe was initially in a false vacuum. The false vacuum caused a rapid acceleration of the Universe, which transitioned into the true vacuum state. The expansion of space was superluminal; therefore, a region that transformed into the real vacuum was unable to communicate

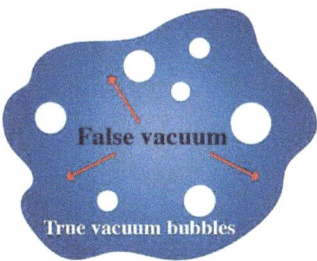

Figure 10.2. Eternal inflation. The area coloured blue is a portion of the Universe in inflationary expansion. The white areas represent bubbles of true vacuum and are universes not in contact and with different physical characteristics.

its state with neighbouring ones. So, if a region close to ours was transformed into a region of real vacuum, given that the expansion of space is superluminal, the two regions would have no knowledge of each other and would be randomly disconnected, constituting two *bubble universes* (Figure 10.2).

In the inflationary phase, and in particular during the passage of the field from the false vacuum to the true vacuum, a "bubble" of the true vacuum (white area) is formed in the false vacuum (blue region) (Figure 10.2).

Since the field reaches a random point of the potential energy minimum in each region, the value of the vacuum energy will be different. Many bubbles of true vacuum will therefore be formed, not just one, as shown in Figure 10.2. Each bubble corresponds to a universe. In other words, inflation ends at different times at different points, and it is possible that in some regions it repeats itself, generating an infinity of universes – a multiverse. The bubbles that have formed expand at a frenetic pace, at the speed of light. Despite this, they never manage to fill the space because the latter expands at a higher speed.

Although an enormous number of bubbles, i.e. universes, are created, if there is only one state of false vacuum and one of true vacuum, we find ourselves in the condition of the quantum multiverse: a large number of universes with the same physical laws. In chaotic inflation, universes exist in a large number of false vacua, and each of them gives rise to different laws of physics, as shown in Figure 10.3 by the different colours of each bubble. Each of these bubble universes will be at enormous distances from the others and will not be able to communicate with them.

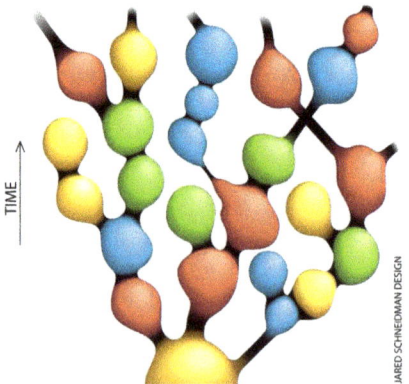

TIME

JARED SCHNEIDMAN DESIGN

Figure 10.3. Formation of universes in bubble theory. Each universe has different physical laws, as shown by the different colours.

Source: Jared Schneidman, *Scientific American*.

The Multiverse of String Theory

One of the fundamental goals of modern physics is the search for a *theory of everything* (TOE) that should bring all interactions into a single framework. Although it is possible that it does not exist, there is still a sort of aesthetic sense that pushes us to seek such a unified description of all phenomena. This quest was begun by Einstein but remained unresolved. Subsequently, it was taken up by several generations of physicists. In 1919, four years after the publication of general relativity, Theodor Kaluza proposed a theory of unification of electromagnetism and general relativity in a five-dimensional space, made up of four spatial dimensions and one temporal one. Kaluza sent the study to Einstein, who initially did not understand the role of the fourth spatial dimension, but then did his best to present it first in 1918 to the Berlin Academy of Sciences and then to have it published in 1921. In 1926, Klein expanded the study by introducing quantum concepts, managing to explain the quantisation of the electric charge based on the quantisation of the momentum in the extra dimension and also calculating the order of magnitude of the extra dimension, which must have been around 10^{-22} m. After an initial interest in the Kaluza–Klein theory, it was almost forgotten until the 1970s. A descendant of the Kaluza–Klein theory in 10-dimensional space is the superstring theory, which is proposed as a possible TOE in which particles are generated by the vibrations of tiny strings of 10^{-35} m. Initially, string theory included only bosons in 26 dimensions and was a theory of strong interactions. Subsequently, Ramond, Neveu, and Schwarz showed how to introduce fermions. The theory's connection with supersymmetry led to *superstring theory*. *Supersymmetry* is a theory according to which there are twice as many known particles.

According to supersymmetry, each boson corresponds to a fermion and vice versa. Furthermore, in 1974, Scherk and Schwarz pointed out that this, rather than a theory of strong interactions, was a *theory of quantum gravity* and, more generally, a TOE. There are five different superstring theories, and they can be considered as five different aspects of a single theory called M theory in an 11-dimensional space. M theory was introduced by Edward Witten in 1995 in the so-called *second string revolution*. The curious thing is that it is not known what the M stands for because Witten never revealed it. Another interesting point is that, as noted by Joseph Polchinsky, string theory also predicts and requires for its consistency the existence of objects of larger dimensions, such as D-branes. For example, a 2-brane is a membrane, a 0-brane is a particle, while a 1-brane is a string. A problem with the theory is that it works in 10-dimensional spaces (nine spatial dimensions plus one temporal), and our world has at most four dimensions (three spatial and one temporal). Where are the dimensions we don't see? As Klein had already explained, these extra dimensions are compact, closed in on themselves, and extremely small in size. In superstring theory, the extra dimensions are rolled up into complex figures called Calabi–Yau spaces (Figure 10.4), six-dimensional spaces named after the Italian Eugenio Calabi and the Chinese Shing-Tung Yau and which are associated at every space-time point. Figure 10.4 gives an idea of what that space is: approximate and distorted because we are visualising a six-dimensional space on a two-dimensional sheet of paper.

The number of possible "rollings" of the extra dimensions is, according to a 2004 estimate, of the order of 10^{500} or, according to other authors, even much greater. There is a Calabi–Yau space for each compactification, and each of these spaces corresponds to a false vacuum (Universe) of the theory. Each false vacuum corresponds to a Universe with physical laws different from those in others. The number of possible universes could be between 10^{500} and $10^{10^{375}}$, the latter value was obtained by Linde. The collection of all possible false voids or universes is usually called a *landscape*. In other words, superstring theory predicts the

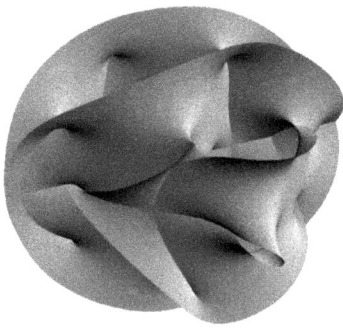

Figure 10.4.   Calabi–Yau spaces.
Source: Wikipedia.

existence of an enormous number of universes, similar to the theory of chaotic inflation.

Having reached this point, one might ask whether superstring theory or rather M theory is a valid candidate for being a TOE. The answer is probably no, because it is not complete and its basic aspects are only now beginning to be understood. Superstring theory had many followers in its early stages and raised hopes that we were close to solving the age-old problem of the unification of gravity and quantum mechanics. Today, however, there are many detractors of the theory. Time has shown that things are not so trivial. There is little or no evidence that string theory is even a solid scientific theory. It predicts supersymmetry, which unfortunately was not observed at CERN. It explains Hawking radiation, but this has never been observed. If the proton decayed, this would be evidence for supersymmetry, but, as discussed in the following, the proton does not decay, at least for 10^{34} years. Supersymmetry predicts dark matter, but this has also never been detected. Yet another predictions of the theory is the existence of new long-range weak forces, which were also never observed. Another fundamental problem is linked to the number of compactifications, which is extremely large, of the order of 10^{500}. Normally, when you test a theory, you try to see whether an observation disproves it or not. In the case of superstring theory, no experiment could disprove it because we could use another compactification and other parameters. This gives rise to a chameleonic theory that can adapt to any result. A theory so flexible that it is unable to predict any phenomenon. However, the "string enthusiasts" do not lose hope, and the research continues.

Is the Multiverse Scientific?

Many physicists do not calmly accept the possibility of the existence of the multiverse, even if some theories, such as superstring theory or inflation, lead to the multiverse. Indeed, much of the criticism of the multiverse comes from researchers working with superstrings, even though this theory predicts the multiverse. An example is Paul Steinhardt, who considers the idea to be dangerous, and as already mentioned, despite being one of the fathers of inflation, he promoted a new model of the Universe because he did not accept the inflationary multiverse. Other people, such as David Gross, criticise the multiverse for other reasons. Physicist John Polkinghorne is one of the detractors of the idea of the multiverse. He writes:

> *Let's recognize these conjectures for what they are. They are not physics, but in the strict sense metaphysics.... By definition these other worlds are unknowable to us. An explanation of equal intellectual respectability ... would be that this world is as it is because it is the creation of the will of the Creator who intends to make it so.*

For the philosopher Richard Swinburne, it is the *culmination of irrationality*.

At this point, the question arises: is the idea of a multiverse a scientific idea? As we have seen, in the case of *chaotic inflation*, the universes are so far away that there is no way to think of directly verifying their existence. The space-time between these universes expands at a speed faster than the speed of light, and since we cannot move at that speed, we have no way to directly verify the existence of these universes, not taking into account the fact that chaotic inflation would also act on our bodies, making them expand and destroying us.

Now, in Karl Popper's view, anything that is not falsifiable is not scientific; therefore, we should think of the multiverse as a non-scientific idea. Wanting to take the side of the supporters of the multiverse, one could think of indirect ways of testing the existence of these universes, and if the tests gave positive results, the idea could be considered scientific. Sometimes, you can have faith in an unverifiable idea of a theory that, as a whole, has been verified and is correct. Let's consider the case of a well-known and verified theory, the theory that describes the strong force, called *quantum chromodynamics* (QCD). This theory predicts that protons and neutrons are made up of more elementary particles: quarks. It also says that it is not possible to observe quarks. From this, we should deduce that QCD is not a scientific theory, but in reality it is, as it makes many other testable predictions, and many experiments have shown that the predictions are correct. Therefore, QCD is not only scientific but also correct. Similar observations can be made regarding general relativity, which can be applied inside black holes, an unobservable region.

Does the Multiverse Exist?

There are some very eccentric proposals to indirectly prove the existence of the multiverse. One of these is based on the very unlikely collision of a universe with ours. In fact, we have already seen that the Universes are at enormous distances from each other. This would leave some sort of sign on the background radiation, but the type of signal would depend on the type of inflation. It could also happen that the push from the collision of the universe close to ours would produce a motion of the galaxies in that area different from that in the rest of the Universe. Some scientists say they have observed this type of *dark flow*, but there is considerable scepticism about this point. In the background radiation, an anomaly called "cold spot" of approximately 5–10 degrees is observed. This area could be due to a variety of causes, such as a large void located between us and the cold spot or other, more exotic, reasons. In an imaginative interpretation, according to Laura Mersin-Houghton, the cold spot could be due to quantum entanglement, but in this case, it would be an entanglement between our universe and another, which were later separated by inflation. Another way the multiverse theory could be tested is by observing the value of constants in our universe. Experiments have been carried out using quasars to study the fine structure of spectral lines and determine variations in physical constants. There don't seem to be any changes in them. Another study is that relating to a uranium deposit in Gabon, which in about two billion years became a sort of reactor. Also, in this case, no variations in the constants are observed. Another way to indirectly establish whether the multiverse exists is to verify the correctness of the theories that propose it, namely inflation and string theory.

As for string theory, as discussed in the section "The Multiverse of String Theory", there is no evidence that it is a good theory. As for inflation, the "smoking gun" of its existence has not been found: primordial gravitational waves. Furthermore, nothing is known about the field that generated it.

Ultimately, the physics community is divided into supporters of the multiverse and its opponents. The idea cannot be directly verified, and it is not trivial to find indirect evidence of its existence. The indirect checks I mentioned are highly speculative. Although there are supporters of the multiverse, even among them the idea is clear: it is a "theory" that is difficult to verify.

Chapter 11

What is Life? Are We Alone?

Ex parvis saepe magnarum momenta rerum pendent.

— Tito Livio

How did life begin? Where do we come from, and who are we? These are the biggest questions we can ask ourselves, and they have certainly always fascinated human beings. Giving an answer poses formidably difficult problems. A large number of scientists, over the past century, have attacked the problem from different points of view, but no one has so far managed to solve it. Despite this, research has made considerable progress, and we hope to be able to provide a solution to the problem in the future. From a general point of view, some answers can be given. The best known is that life was created by a deity. As we know, there are and have existed a large number of ideas on the origin of the world linked to divinities. These beliefs have always been proposed as science without there being any explanation of the creator God. In more recent times, a more cryptic theory has appeared: *intelligent design*, which does not deny the reality of evolution but maintains that the extreme complexity of living beings can only be explained by the existence of an intelligence that directed the evolution to follow the paths he followed. The second follows the astrophysicist Fred Hoyle and his *theory of the stationary universe*, according to which the Universe would be eternal, life would never have had an origin, and life would always have existed. His collaborator, Wickramasinghe, also supported this thesis and the idea that space was full of life (viruses, bacteria, etc.). With the discovery that the theory of the stationary universe was wrong, the idea that life had always existed also decayed. A third possibility is that life did not originate on Earth but arrived from space. This is the *panspermia theory*. Proponents of this theory are Svante Arrhenius, Fred Hoyle, and Francis Crick, among many others. According to Crick, there would have been too little time for a terrestrial origin of life, and according to him, it was more likely that an extraterrestrial civilisation spread life everywhere. This is the *guided panspermia* hypothesis. The idea of panspermia obviously only shifts the problem of the birth of life from our planet to another place but does not solve it. Panspermia is a scientific theory, but it is considered less probable than the local origin of life, i.e. on Earth. Today, we still don't know how things happened, but step by step, we are getting closer to knowing how they could have happened. Despite the progress made in the field, the controversy still exists between the idea that life is the result of *chance*, however extraordinary, according to the ideas of Jacques Monod, or of a *necessity* imposed by natural laws, as supported by Ilya Prigogine. The study of the origin of life is so complex that it requires interdisciplinary work. Nobel Prize winners in physics, chemistry, and biology have dealt with and continue to deal with this complex and extraordinarily interesting problem.

Some Ideas on the Birth of Life: The Primordial Soup

Giving a definition of life is not trivial, and it is even less trivial to talk about its origin. Despite this, Darwin had his own ideas on the origin of life, which he did not publish but spoke about in his private correspondence. In an 1871 letter addressed to his friend Hooker, Darwin spoke of the origin of life, starting with chemical processes fuelled by energy sources. In the letter, he spoke of a "small warm pond" as a possible primordial soup in which the first living organisms would have formed. In his words:

> *If we could conceive of a small, warm pond, containing ammonia and phosphoric salts, light, heat, electricity, etc., so that a protein was chemically produced ready to undergo new and more complex changes...*

And a few years later, in 1882, to Daniel Mackintosh, he wrote:

Although there is still no evidence in favor of the hypothesis that a living being developed from inorganic matter, I cannot help but believe in the possibility that this will one day be proven.

Only in the 1920s was the problem of the origin of life taken up again by the Russian biochemist Alexander Ivanovich Oparin and the English geneticist John Haldane. Both conceived the idea of *chemical evolution*, that is, the idea that in the primordial seas, there existed an *organic soup* which, increasing in complexity, would lead to the formation of simple cells, the point of origin of all living beings. Due to the poor development of analytical chemistry for many years, there were no advancements or experimental ideas to verify the ideas of Oparin and Haldane. The English version of Oparin's second book was read by Harold Urey, winner of the Nobel Prize for Chemistry in 1934. In a seminar held at the University of Chicago on the origin of the solar system and, in particular, on experiments in the formation of organic compounds, there was present a recent graduate in chemistry, Stanley Lloyd Miller. Some time later, Miller showed up in Urey's studio with a proposal for implementing some experiments. Urey agreed to do the experiments, but if there were no positive results in six months, they would change Miller's thesis. Together, they built a device containing liquid water and the gases hydrogen, ammonia, and methane. This was the idea at the time about the constitution of the Earth's atmosphere. The water was boiled in a pipette at the bottom and then cooled to form what was rain falling in the primordial oceans. Electric discharges of 6,000 V were produced in the pipe above, representing the lightning of the primordial Earth. After a few days, the water changed colour, becoming reddish-brown. The analysis showed that various organic compounds had formed, including *amino acids* (glycine and alanine), which, as already mentioned, are the "building blocks" that form proteins.

In Short: Nucleic Acids: DNA and RNA

Nucleic acids (RNA and DNA) are macromolecules responsible for the conservation and transmission of genetic information. DNA has a double helix structure that resembles a ladder. RNA, on the other hand, has only

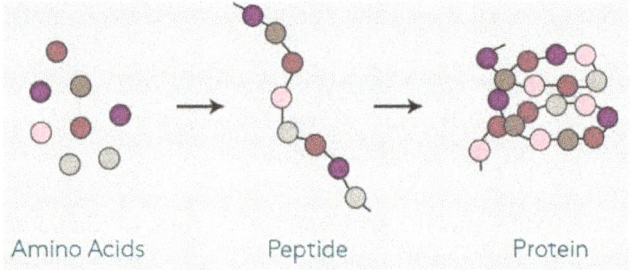

Figure 11.1.   Amino acids and proteins.

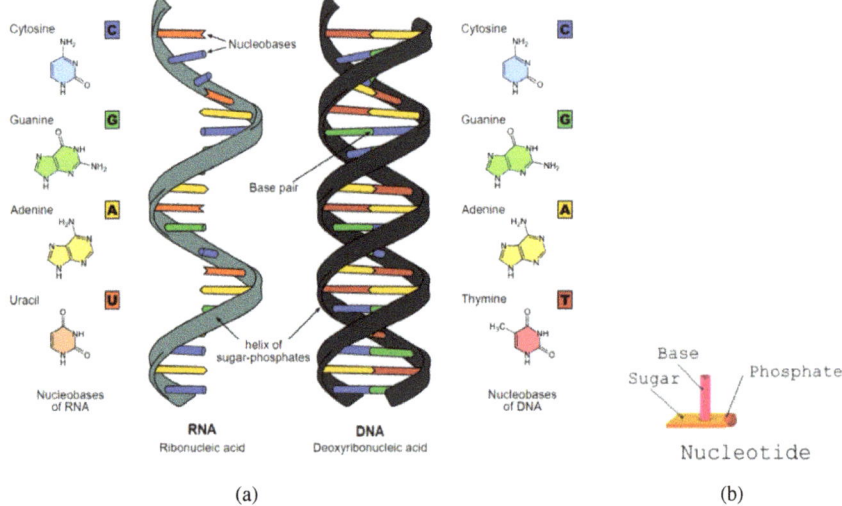

Figure 11.2.   (a) Nucleic acids: RNA and DNA. (b) Nucleotide.

Source: Reproduced from https://it.wikipedia.org/wiki/Acidi_nucleici; https://www.my-personaltrainer.it/biologia/nucleotidi.html.

one helix. The small segments connected to the helices are the *nitroge-nous bases* (A, G, C, T, U). The helix is made of sugar phosphates. The simplest units capable of forming a chain of RNA or DNA are called *nucleotides* (see Figure 11.2(b)). Nucleotides are made up of a sugar called *ribose*, combined with a phosphate group from which an oxygen atom has been removed, and a base. The union of many nucleotides con-stitutes a chain (called a *polynucleotide chain*), where the phosphate groups and sugars form the helix from which the nitrogenous bases protrude. The bases are matched according to a precise rule. In the case of DNA, A is paired with T and C with G. So, if one strand of the double helix begins with AGGTCCGTAATG, the other will be TCCAGGCATTAC. That is, by knowing one filament, you can deduce the other. The sequence of bases carries a message with this four-letter alphabet that conveys information for a protein. Furthermore, each group of three nucleotides, called a *codon*, encodes an amino acid. Since there are four *nucleotides*, there are 4^3 possible triplets available to encode the 20 amino acids. Genes are segments of DNA that contain the code for a specific protein. A complete set of genes makes up the genome. The human *genome* is 3.5 billion letters long. In the case of RNA, G pairs with C and A with U.

The results of the experiment gave considerable support to the idea of *chemical evolution*. The experiment generated great expectations and gave a strong impetus to carry out other experiments. An important point to remember is the fact that the atmosphere used by Urey and Miller was a *reducing* atmosphere, while today it is believed that it had to be more *oxidising*, i.e. it tends to lose electrons in a reaction containing CO_2 and water. By repeating the Urey–Miller experiment with an atmosphere of this type, the reaction yield is much lower, i.e. not all the substances in the Urey–Miller experiment were formed. Despite this mistake, the Urey–Miller experiment gave birth to *prebiotic chemistry*, stimulating the performance of a multitude of other experiments. Of notable importance are the results obtained by the Spaniard Joan Orò. Until the late 1950s, the *bases of nucleic acids*, namely the chemicals adenine (A), guanine (G), cytosine (C), thymine (T), and uracil (U), had not been found. Figure 11.2(a) shows the location of the bases in the structure of RNA and DNA. In 1959, Orò conducted an experiment starting with hydrocyanic acid (HCN) to obtain a base of nucleic acids, adenine (A), and together with Miller, they found guanine (G), while all the other bases were found by other researchers. Thousands of experiments were carried out to simulate the early Earth using different forms of energy (ultraviolet radiation, visible radiation, heat, radioactivity, or electric discharges) in different environments (aquatic, gaseous, or atmosphere–water interface). Most of the experiments highlighted the fundamental role of hydrogen cyanide and formaldehyde (HCOH). In 1969, the Murchinson meteorite fell, which takes its name from the place in Australia where it was found. It is a *carbonaceous chondrite* containing 14,000 different organic compounds, including 70 amino acids, but it is estimated that there could be millions of them. In addition to amino acids, the *bases of nucleic acids* have been found. These substances are not due to contamination, as proven by the presence of organic compounds that are not found on Earth. The composition is similar to that of the results of the Urey–Miller experiment. This finding is of considerable importance because it shows that, although due to the low-reducing character of the Earth's atmosphere, the primordial soup did not produce many compounds (as in the Urey–Miller experiment in a low-reducing environment). The contribution of extraterrestrial material could play a fundamental role in the birth of life. The study of comet *67 P/Churymov-Gerasimenko* a few years ago showed the existence of HCN, nitrogenous compounds, aldehydes, and alcohols. So, comets could also provide a certain amount of probiotic molecules.

From the Primordial Soup to LUCA

However, even assuming that chemical processes on Earth or in space could generate amino acids, sugars, nucleotides, etc., from these substances to the formation of life and the so-called last universal common ancestor (LUCA), there is a long way to go. A *metabolism* is needed for the functioning of the ancestral organism. Metabolism is the ability to extract energy from the surrounding medium and use it to stay alive. Metabolism is based on proteins that are fundamental to the life process, *enzymes*. These enzymes are synthesised starting with the information present in nucleic acids. Unfortunately, for nucleic acids to duplicate and express information, enzymes are required. We find ourselves faced with the chicken-and-egg problem in the field of the origin of life: *enzymes are produced by nucleic acids, but for the latter to duplicate, enzymes are needed*. It would be necessary for the components of life to form all together and collaborate to generate it.

Leslie Orgel set out to simplify the problem and suggested that primordial life had no proteins or DNA. The engine of life was composed almost entirely of RNA. For this to work, the primordial RNA molecules had to be very versatile, and first, they had to be able to make copies of themselves. The idea that life began with RNA proved to be very influential. RNA can do something that DNA, a rigid double-stranded structure (double helix) (see Figure 11.2) cannot. As a single-stranded molecule, it could fold into a variety of shapes, and such folds seemed similar to the way proteins, which are long strands of amino acids rather than nucleotides, behave. If RNA could fold like a protein, perhaps it could form enzymes. If this were true, RNA would be capable of storing information and, at the same time, catalysing reactions like enzymes do. Thomas Cech, in 1982, and Sidney Altman, in 1983, showed that these ideas made sense, showing that some RNAs have catalytic capabilities (like enzymes) and that they can function as enzymes. Now, the idea that life began with RNA seemed promising. The discovery has big implications for our discussion of the formation of life. In fact, it is no longer necessary that the proteins and the nucleic acids that encode them are formed at the same time. This leads to the idea of the so-called *RNA world*, a term coined by Walter Gilbert. According to Gilbert, the first stage of evolution consisted of "RNA molecules carrying out the catalytic activities necessary to assemble themselves from a soup of nucleotides". In such a world, RNAs would be capable of doing the things important to the formation of life, i.e. carry genetic information and function as catalysts for chemical reactions.

The RNA world is an elegant way of reproducing the complexity of life from scratch. Instead of relying on the simultaneous formation of large numbers of biological molecules from the primordial soup, some sort of "jack-of-all-trades molecule" could do the job of all of them. Thus, the discovery of the enzymatic capabilities of RNAs implies that the RNA world had a complex metabolism. The organisms of this world would undergo evolution through natural selection. Although the existence of organisms capable of evolving is confirmed, some open points remain, such as the problem of the invention of the genetic code and the synthesis of proteins. Furthermore, the experiments are a long way from producing RNA. The problem remains of how it could have formed on the primordial Earth. Gerald F. Joyce and Leslie Orgel argued that the spontaneous appearance of RNA chains on the early Earth "would have been something of a miracle". The discovery of catalytic RNA led to the idea that it would account for at least two of the fundamental aspects of life: the genetic aspect and metabolism. The problem of explaining how it appeared remains. As we have already said, Oró showed that the nucleotide bases (A, G, etc.) can be obtained in particular reactions, but a complex question remains unanswered: How to generate an *RNA nucleotide* (called a *ribonucleotide*) by connecting each base with the sugar (known as D-ribose), which makes up RNA, and this with a phosphate group (see Figure 11.2(b)). Sugar (D-ribose) is formed in experiments; the source of phosphate could be polyphosphates present in volcanic emissions or meteoric phosphates. Despite various attempts to obtain the necessary unions, it was not possible to carry them out. The other problem is that even if we manage to build nucleotides, we must then bind them to form chains, or polymers. Furthermore, apart from the problem of RNA synthesis, its existence does not ensure the existence of an RNA world since it must be able to replicate. Several researchers continued to search for a way to replicate RNA. Despite all attempts, the problem has not been resolved. The lack of self-replicating RNA is a fatal problem for the idea of the RNA world. RNA does not appear to be capable of giving rise to life.

Other Pathways: The Iron–Sulphur World Hypothesis and Metabolism

The RNA world theory is based on the idea that the most important thing for an organism is to reproduce. However, there are many researchers who do not believe that reproduction is fundamental. Before reproducing, an organism must be self-sustaining. To stay alive, you need to absorb some form of energy. So, they think the starting point is *metabolism*. As already mentioned, we call metabolism the ability to extract energy from the surrounding medium and use it to stay alive. This process is so important that many researchers think it is the first thing life ever did. There is therefore a research chain that is based on the idea that metabolism comes first. What do organisms that only have metabolism look like? One of the most interesting ideas is that of Günter Wächtershäuser from the late 1980s. For Wächtershäuser, the first organisms were completely different from everything we know and were not made of cells; they would have been *acellular* and had no enzymes, DNA, or RNA. Both nucleic acids and information-carrying molecules were missing. However, they would have had a certain metabolism that developed in two dimensions, not three, and a capacity for evolution. These compounds, like the terminal or intermediate products of metabolism (metabolites), had a negative charge and were anions. Wächtershäuser imagined a stream of hot water, rich in volcanic gases such as ammonia and traces of volcanic minerals, flowing from a volcano. Chemical reactions began to occur where water flowed over rocks. Metabolic cycles were created, i.e. processes in which one chemical substance is converted into others, until the initial substance is recreated. In the process, the system absorbs the energy used to start the cycle. These metabolic cycles were nothing like life. Wächtershäuser spoke of *precursor organisms* that could barely be called living. He developed his model in the 1980s and 1990s in great detail, outlining which minerals were in play and the chemical cycles that took place. It was a theory that

Figure 11.3.   Black fumaroles.

needed a discovery to support its ideas. This discovery had previously been made in 1977 by a team led by Jack Corliss using a submarine near the Galapagos Islands. Corliss and collaborators observed volcanically active rock ridges rising from the sea. These ridges were covered in hot springs. These areas were populated by a multitude of different types of animals. In other words, the scenario of this model and in which life was formed is that of underwater hydrothermal sources (hydrothermal vents known as *black fumaroles*) at the bottom of the oceans (Figure 11.3).

In the model, all organic compounds were formed *in situ*. The energy and "reducing power" necessary to transform CO and CO_2 into organic matter would be due to the reaction of the formation of pyrite (FeS_2), starting with hydrogen sulphide (H_2S) and iron sulphide (FeS). The confirmation of this thesis came experimentally in 1990. After this, Wächtershäuser proposed a whole series of reactions that started with the assimilation of CO and CO_2 and ended with the generation of cells. Furthermore, current metabolic pathways would have been preceded by a series of reactions not catalysed by enzymes. As a first step, we must show how the incorporation of inorganic carbon occurs. According to the author, this was possible thanks to an autocatalytic CO_2 fixing process. This cycle would promote the fixation of CO_2 into organic molecules. Wächtershäuser called his hypothesis the *iron–sulphur world*. Geologist Mike Russell found fossil evidence of thermal chimneys with temperatures below 150°C in the 1980s. Furthermore, the fossil remains of these chimneys contained pyrite, and he suggested that the first complex organic molecules formed within the pyrite structures. Russell also suggested that thermal vents in the deep sea, warm enough to allow pyrite structures to form, harboured Wächtershäuser's organisms. According to this thesis, life began at the bottom of the sea, and metabolism appeared first. The idea was further modified based on a discovery by Peter Mitchell: the *Mitchell proton gradient*.[1] It was clear that a proton gradient is needed to store energy. Starting with this idea, Russell concluded that life forms where a natural proton gradient exists. The ideal place would be a hydrothermal vent with proton-poor water, i.e. alkaline water. So, the acidic Corliss vents, in addition to being too hot, would not have worked for this purpose. The first alkaline hydrothermal vents were discovered by Deborah Kelley in the Atlantic in a place that was called "Lost City". The water temperature is between 40°C and 75°C and slightly alkaline. These vents were perfect for Russell's ideas (who, in the meantime, began collaborating with the biologist William Martin), who became convinced that these were in reality the places where life was born. The rocks of the vents are porous and formed sort of pockets containing pyrite, among various chemical substances. Combined with the natural proton gradient from the vent, these were ideal locations for metabolism to begin. After life had exploited the chemical energy of the chimney water, the production of molecules such as RNA began. With the formation of the membrane, a true cell is formed, which then goes from the porous rock to the open sea.

In the past decade, a third approach has appeared that promises a way to create an entire cell from scratch.

Cells from the Beginning: The World of Lipids

In the RNA world, membranes were dispensed with, although it is clear that it is unlikely that RNAs would be found in solution without protection. In Wächtershäuser's hypothesis, even if the membranes appeared,

---

[1] Change in the value of a quantity (such as temperature, pressure, or concentration) with change in a given variable and especially per unit distance in a specified direction.

they did so at a later stage. Researchers have become aware that it is difficult to imagine life forms without membranes. Michael Russell underlines the importance of iron–sulphur membranes formed in a not-very-hot and alkaline environment.

Membranes are important in cell energetics and proton gradients. In today's cells, membranes are made up of lipids and proteins. Lipids are the essential element for closing vesicles because they are molecules that have a polar [2] and a non-polar part. They can associate with each other and self-organise through the non-polar zones, placing the polar zones in an aqueous medium that is non-polar. The membranes grow, and the vesicles swell, and at a certain point, they divide spontaneously. In addition to the ease of formation and the ability to form microenvironments, vesicles can generate proton gradients, acting as a barrier between two apolar regions. As shown by Deamer in 2015, through hydration–rehydration cycles, lipid vesicles give rise to the polymerisation of nucleotides. During rehydration, the RNAs are incorporated into the vesicles. Previous results have led several researchers to propose a *lipid world* [3] that would precede the RNA world. Jack Szostack set himself the goal of achieving RNA replication together with the growth and reproduction of vesicles, i.e. an RNA cell capable of evolving.

This last story is linked to a collaboration between Szostack and the champion of the idea of "compartmentalisation first", namely Pier Luigi Luisi. The ideas of the latter can be traced back to those of Oparin's coacervates. Luisi's challenge was to create protocells, but despite various experiments, he failed to create anything truly realistic and convincing. In 1994, he suggested that the first protocells must contain RNA, which must be able to replicate within the protocell. This idea quickly gained a supporter: Jack Szostak. Szostak's team managed to build protocells that retain their genes and simultaneously absorb molecules from the outside. Protocells can grow and divide, and RNA can replicate inside. These results and those of Sutherland, already mentioned, suggest a new unified approach to the origin of life, based on which all three functions, which we have indicated several times, can be achieved simultaneously.

---

[2] Polarity is a property of molecules whereby a molecule has a partial positive charge on one side of the molecule and a partial negative charge on the opposite side of it.

[3] Substances present in animal and plant cells and tissues characterised by insolubility in water and solubility in organic solvents.

Order and Life

What distinguishes the living from the non-living is the order based on information. According to Ilya Prigogine, the appearance of living organisms is not an accidental event but is implicit in the irreversible processes of systems far from equilibrium. There would be a relationship between processes of spontaneous self-organisation and the birth of life. There is a kind of plot that connects the non-living with the living. Matter is structured so as to become living matter.

Understanding the origin of life is a puzzle of extreme difficulty. In half a century, several steps forward have been made in its solution. Today, we know many mechanisms that were even more difficult to imagine; we know better than several decades ago about where we come from, but we certainly haven't reached our goal. Despite the thousands of experiments and decades spent trying to reproduce the formation of life on the primordial Earth, to date, no success has been achieved. We have obtained amino acids and nucleotides, and theoretically, in the laboratory, we could create the macromolecules at the basis of life; however, in practice, we have not clarified the origin of life on Earth. Life is made up of long chains of amino acids, nucleic acids, and so on, but these chains are ordered. The sequences of amino acids and nucleotides that follow a specific order. If we use all 20 amino acids to form a chain of 50 randomly organised amino acids, we can obtain an enormous number of chains, almost as many as all the atoms in the Universe, all different from each other. Unlike this situation, life is based on order and, more precisely, as we will see, on the order linked to information. For example, proteins are long chains with a precise order of amino acids. The same order dominates in genes and nucleic acids. The question that comes naturally is: How was this order established? If we knew this, we would have already solved our puzzle. The typical order of life can take us down two different paths. That of the religious person who supposes that there is an entity that has infused rules into matter that push it towards order. There would be an organising being that would make matter pass from states of greater disorder to the order necessary for life to form. The other path is that of science, and today it tells us, in the voice of some of its exponents, such as Ilya Prigogine, that the appearance of living organisms is not an accidental event but is implicit in the irreversible processes of systems far from equilibrium. There would be a relationship between processes of spontaneous self-organisation and the birth of life. It is as if there is some sort of necessity in the world of non-life that pushes it in the direction of the living. Disorder does not constitute the rule for matter but only an intermediate stage that moves in the direction of creating ever lower disorder until order is achieved. Prigogine's ideas are based on one of his inventions, *dissipative structures*, i.e. thermodynamically open systems that exchange energy, matter, and entropy with the environment and are far from equilibrium. Open systems with *autocatalysis*,[4] moving away from equilibrium, amplify small fluctuations and, as a final result, self-organise while dissipating entropy. In *The New Alliance*, Prigogine and Isabelle Stengers give an example of how "the instability of a stationary state gives rise to a phenomenon of spontaneous self-organization". It's about Bènard's instability. A liquid is placed in a container, and the lower surface of the horizontal liquid layer is heated to a higher temperature than that of the upper layer. A flow is generated from bottom to top, and when the gradient exceeds a threshold value, the fluid becomes unstable. A complex structure, consisting of hexagonal structures, is generated on the upper surface. A similar thing happens in dissipative structures. While in the Bènard cell the instability has mechanical origins and the behaviour of the fluid flow is predictable, in chemical systems the situation is different. As reported by the two authors:

> *The fate of the fluctuations that perturb a chemical system, as well as the regime of new situations towards which it can evolve, depends on the detailed mechanism of chemical reactions. In contrast to situations close to equilibrium, the behavior of a system far from equilibrium becomes highly specific. There are no longer universally valid laws from which the general behavior of the system could be deduced, for each value of the boundary conditions. Each system is a case in itself, each set of chemical reactions must be explored and can produce qualitatively different behavior.*

Dissipative structures constitute a source for explaining life, not as an epiphenomenon due to chance but as one that arises from the same laws of nature and matter without excluding chance. There is a kind of plot that connects the non-living with the living. Matter is structured so as to

---

[4]Autocatalysis is a catalytic process in which the catalyst is represented by one of the reaction products or intermediates themselves.

become living matter. There is order beneath the chaos, which led Prigogine to say that:

Each molecule knows what the other molecules will do simultaneously with it and at macroscopic distances [...] molecules communicate [...] [This property] in non-living systems comes at least unexpectedly.

Like religion, Prigogine's science also leads us to the conclusion that there is a tendency for matter to always organise itself towards higher-order stages, even to the point of giving birth to life. Without this tendency towards order, it would be difficult to explain the formation of life in reasonable time spans. For example, the formation of a single RNA molecule from nucleotides, randomly, would take much longer than the age of the Universe. However, several scholars believe that Prigogine's self-organisation is actually not sufficient to explain the transition from non-living to living.

The complexity of the origin of life without there being an underlying order led Francis Crick to conclude that:

The mechanism necessary to make the genetic code operational, which is universal, is too complex to have arisen in one fell swoop. Any man with his wealth of knowledge at our disposal today could only affirm that the origin of life seems to belong to the order of a miracle in its present state, so many conditions would have to be brought together in order to achieve it.

The problem of the origin of life is so complex that it leads scientists to speak of a miracle. This is completely unexpected and leads us to ask ourselves whether it is a mere expression to convey the idea of complexity or whether the religious sphere is truly being brought into play.

Are We Alone?

In 1960, Frank Drake used the Green Bank radio telescope in an attempt to find extraterrestrial life by studying radio waves emitted by hydrogen atoms. He pointed the radio telescope at the star Tau Ceti and then Epsilon Eridani. Initially, he thought he had received a signal of extraterrestrial origin, which then disappeared and reappeared a few days later. It was clear that it was interference coming from Earth. Drake's search never stopped, but it never produced positive results. Drake himself, in an attempt to estimate the number of extraterrestrial civilisations in the Milky Way, wrote an equation consisting of the product of various terms:

$$N = R^* f_p n_e f_l f_i f_c L$$

where:

- N is the number of extraterrestrial civilisations present today in our galaxy with which we can think of establishing communication;
- R^* is the average annual rate at which new stars are formed in our galaxy;
- f_p is the fraction of stars that have planets;
- n_e is the average number of planets capable of hosting life forms;
- f_l is the fraction of planets on which life has actually developed;
- f_i is the fraction of planets on which intelligent beings have evolved;
- f_c is the fraction of extraterrestrial civilisations that have developed a technology;
- L is the time span in which civilisations can transmit signals that can be picked up on Earth.

To know the number of extraterrestrial civilisations, you need to know the values of all the terms in the equation. Drake and collaborators gave their estimates of the various parameters: $R^* = 10$ stars per year; $f_p = 0.5$, assuming that half of the stars have planets; $n_e = 2$, i.e. each planetary system has two planets that can support life; $f_l = 1$; f_i and $f_c = 0.01$, taking the Earth as a model; $L = 10,000$. This gives a value of $N = 10$. This estimate is based on a deep ignorance of the terms in Drake's equation. In Drake's time, it wasn't even known if extrasolar planets existed. The most recent findings have led to new estimates for various parameters.

In the Drake equation, the terms R^*, f_p, and n_e are obtained from astrophysics and are quite well known today. The terms f_l, f_i, and f_c are related to biology and are therefore almost unknown. The last term is linked to the lifespan of an advanced civilisation, and we still don't know how much its value could be. As mentioned, from 1960 to today, much progress has been made, especially in the determination of the parameters of astrophysical origin in the Drake equation, which we now summarise. As for R^*, calculations by ESA and NASA suggested that the star formation rate was 7 stars per year, while calculations from 2010 led to a number of

1.5–3 per year. As for f_p, the first extrasolar planets were discovered in 1995 when studying the oscillation produced by the planet's gravity on the star. The method mainly allowed the discovery of Jupiter-type planets. With the Kepler telescope, which is based on another technique, the decrease in brightness due to the concealment of the star by the planet (the transit method), and with other techniques (microlensing, etc.), today we have identified more than 5,000 confirmed extrasolar planets. In general, the study of extrasolar planets has led to the conclusion that about 40% of Sun-like stars have planets, while infrared observations of dust discs around young stars have led to the conclusion that between 20% and 60% of Sun-like stars must contain planets. This leads us to conclude that in our galaxy containing hundreds of billions of stars, there should be billions of planets. Only a small fraction of them are suitable for hosting life. Estimating the third term in the Drake equation is an extremely complex problem. As of December 20, 2023, the number of habitable exoplanets, as published by the *Planetary Habitability Laboratory* of the University of Puerto Rico in Arecibo, would be 63. These are exoplanets with a greater probability of having a rocky composition with a mass lower than 6 Earth masses and which orbit in the "conservative" habitable zone, i.e. that part of the habitable zone where favourable conditions are maintained for a good part of the life of the star in the main sequence, the band of stars arranged in an almost diagonal sense in the Hertzsprung–Russell diagram, which is a graphic representation that relates the effective temperature (reported on the abscissa) and brightness (reported on the ordinate) of the stars. Habitable planets are found in the *habitable zone*, which is the region around a star where a planet would have a temperature such that liquid water is found on the surface. The size and distance from the star of the habitable zone depend on the temperature of the star. In the case of our solar system, the Sun's habitable zone lies between 0.95 and 1.5 times the Earth–Sun distance. Of particular interest are the planets GJ 1002b and GJ1002c, which are similar to our Earth and could host life. They are 16 light years away from us. Both planets are found to rotate around the red dwarf star GJ 1002, a rather cold and faint star and with a mass of only one-eighth of the mass of our Sun. GJ 1002b takes 10 days to complete a rotation around its star, while GJ 1002c takes 20 days. Both exoplanets must be close enough to their stars to be habitable because the star is much dimmer than the Sun.

Another planet that has been considered a habitable planet is Gliese 581c, located around the star Gliese 581. It is located 20 light years from

Earth, its size is one-and-a-half times that of Earth, and its mass is between 5 and 10 times that of Earth. The temperature is the right one to have liquid water on the surface, and initially, it was thought it might be capable of supporting a terrestrial life form, such as *extremophiles*. In reality, the planet always presents the same face to the star; therefore, it was concluded that life would only be possible in a restricted region of the planet, the *terminator area*, i.e. the moving line that divides the daytime and nighttime areas of the planet. A 2013 study based on data from the Kepler telescope concluded that there could be 40 billion Earth-sized planets orbiting in the habitable zones of Sun-like stars and red dwarf stars within the Milky Way, and of these, 8.8 billion may be orbiting Sun-like stars. If you consider that there are about 100 billion stars in the galaxy, the product f_p would be about 0.4. Considering another 2020 study based on the planets discovered by Kepler, if we set the number of stars in the Milky Way to be 400 billion, it would mean that there are 6 billion Earth-like planets in the habitable zones of stars similar to the Sun. Some astronomers have pointed out that assuming that life can only form in the habitable zone could deprive us of the possibility of discovering exotic life forms, which do not require terrestrial conditions for their birth. For example, on Europa, Jupiter's moon, in an area very far from the habitable one, due to the tidal effects produced by Jupiter, there could be liquid water under the frozen surface. The next term in the Drake equation, f_l, is the fraction of planets on which life has actually developed. In 2002, Charles H. Lineweaver and Tamara M. Davis, using statistical reasoning based on the time necessary for the development of life on Earth, estimated the value of f_l. From their study, it would appear that for planets older than a billion years, this value should be greater than 0.13. From the study of the origin of life on Earth, it leads us to think that that f_l could be very high. Life on Earth appears to have begun as soon as favourable conditions arose, a few hundred million years after the Earth had cooled. This has led to the idea that abiogenesis might be relatively common where conditions make it possible. The same idea also seems to be supported by the results of experiments such as the Miller–Urey one, which demonstrate that under appropriate conditions, organic molecules can form spontaneously from simple elements. Scientists have been looking for a proof for this theory by looking for bacteria that are not related to other life forms on Earth, but none have yet been found. Furthermore, these conclusions are based on Earth data and therefore have a very small statistical basis. Francis Crick and Leslie Orgel hypothesised that life on Earth began with microorganisms deliberately sent to Earth by an advanced civilisation on another planet by means of a special unmanned spacecraft on long-range voyages: *guided panspermia*. In 2020, an "astrobiological-Copernican" principle was proposed, based on the principle of mediocrity, hypothesising that intelligent life would have formed on other planets, similar to Earth, in a similar way to what happened on Earth; therefore, within a few billion years, life would form automatically as a natural part of evolution. As part of the study, f_l, f_i, and f_c are all set to a probability of 1 (certainty). If this were true, the resulting calculation would conclude that there are more than 30 technological civilisations currently in our galaxy. From an observational point of view, information on f_l can be obtained from the study of the alterations induced by life forms on the chemical composition of the planet that hosts them. In fact, living beings release methane and oxygen into the atmosphere. The presence of oxygen and methane is an unequivocal signal of the existence of life forms on a planet. We therefore need tools that allow us to collect a sufficient quantity of the light emitted by exoplanets and carry out a spectral analysis of the atmosphere. Until we have concrete data, it will be difficult to understand whether life exists on a planet or not and to obtain an observational value of the term f_l. Even for the other terms in the Drake equation, we cannot go much further than conjectures, given that the only example of a planet on which intelligent life (f_i) capable of developing technology and communicating (f_c) has formed is the Earth. From the fossil findings, it can be deduced that the evolutionary process on Earth started slowly and then accelerated. For more than half of Earth's life, life did not go beyond the form of cells without a nucleus, or *prokaryotic cells*. Only two billion years ago, cells acquired a nucleus and began to work together, giving rise to multicellular organisms. The most complex life forms appeared half a billion years ago with the Cambrian explosion: in a period between 70 and 80 million years, almost all animal groups developed. The value of f_i is particularly controversial. Biologist Ernst Mayr and colleagues highlighted that, of the billions of species on Earth, only one has become intelligent and, therefore, the value of f_i must be low. The opponents of this thesis, who are in favour of a higher value of f_i, underline instead that the complexity of living beings tends to increase along the course of evolution and conclude that the appearance of intelligent life, sooner or later, is almost inevitable. These ideas are criticised by supporters of the *rarity Earth hypothesis*, while the *snowball Earth scenario* or events leading to mass extinctions have highlighted the possibility that life on Earth is relatively fragile. Research into any past life on Mars is relevant since the discovery that life formed on the red planet but ceased to exist increases the f_i estimate but would also indicate that in half of the known cases, intelligent life did not arise and never developed. f_i, according to SETI Institute scholar Pascal Lee, should have a very low value, of the order of 0.0002. Even more speculative is the fraction of extraterrestrial civilisations that have developed technology and are capable of communicating with other possible civilisations. As far as we are concerned, 1971 marked the beginning of the era of the search for intelligent extraterrestrial life (Search for Extraterrestrial Intelligence, or SETI). In that year, NASA commissioned a study to design a large telescope obtained by connecting many telescopes: *Project Cyclops*, which is capable of detecting radio signals from a planet about a thousand light-years away from Earth. Cyclops was never built. Today, there is a programme called "SETI@home", a software that can be installed by anyone who wants on their computer and uses the moments of inactivity of their computer to analyse the data of the SERENDIP project (Search for Extraterrestrial Radio Emissions from Nearby Developed Intelligent Populations) from the University of Berkeley. SERENDIP is a receiver that is part of a radio telescope, receiving signals from all celestial bodies observed by the latter. To date, there have been no signals from extraterrestrial civilisations. Assuming that advanced extraterrestrial civilisations exist, the problem remains that the dimensions of the galaxy (radius of approximately 53,000 light years) are enormous and electromagnetic radiation can only move at the speed of light. Assuming that advanced civilisations exist, they would need to be at distances that were not too great to be able to detect their signals if they decided to reveal their existence by emitting radio waves. The final factor in the Drake equation is the time it takes for an extraterrestrial community to communicate its existence. An estimate of L was made by Michael Shermer based on the average duration of 60 historical civilisations. The value of L obtained is equal to 420 years. This result is based on the observation that, on Earth, technical evolution occurred in a linearly increasing manner as each civilisation preserved the discoveries of the previous ones. There is a form of the Drake equation that takes reappearance factors into account. In other words, even if a civilisation ended after a certain number of years, on the planet, life could continue for very long periods, giving the possibility for another civilisation to evolve. Therefore, different intelligent forms and different civilisations could follow one another on a planet during its existence. This could increase the value of L. Pessimists point out that the ability to transmit signals appeared around the same time as atomic weapons appeared.

Civilisation could therefore be destroyed by the development of technology itself. Some believe that our civilisation cannot survive more than a couple of hundred years of the development of technology. There are opposing points of view. In theory, our civilisation could still exist for a billion years, enough time for the Sun to increase its brightness by 10%, after which our civilisation will not be able to survive. Astrobiologist David Grinspoon has suggested that once a civilisation has developed, it may overcome all threats posed to its survival and then survive indefinitely, taking L to the order of billions of years. The final term in the Drake equation is probably the one that dominates the others. The longer the average lifespan of an advanced civilisation, the greater the number of civilisations populating the galaxy and the greater the probability of intercepting communications. Obviously, each variation of the parameters produces notable changes in the number N. The values proposed by the optimists are around $N = 600,000$, and those of the pessimists are around 0.0000001 (excluding Earth in the calculation). Criticisms of the Drake equation mostly follow from the observation that many terms in the formula are largely completely conjectural. This aspect is clearly expressed by T. J. Nelson:

> *The Drake equation consists of a large number of probabilistic factors multiplied together. Since each factor is definitely between 0 and 1, the result is also a seemingly reasonable number definitely between 0 and 1. Unfortunately, all values are unknown, making the result less than useless.*

Another limitation of the equation consists in the fact that the parameters that appear in the formula refer to life understood in strictly terrestrial terms, that is, to a type of approximately humanoid beings. In principle, it cannot be completely ruled out that intelligent life forms radically different from humans could develop, for example, on Jovian-type planets. By avoiding, as many believe it is appropriate to do, anthropocentric positions, the number of intelligent species in the galaxy could, in theory, increase significantly. Another aspect to note is that the arguments for the existence of life on a planet are closely linked to the fact that the planet must remain within the so-called *Goldilocks* zone. This geometric condition, based on the average radius of the orbit and its eccentricity, is a necessary but not sufficient condition for the existence of life. There are other necessary conditions, such as the existence of a magnetic field that defends the inhabitants from stellar winds. The Drake equation should also take this into account and more. If giant planets, such as Jupiter, were not present in our solar system, life probably would not have formed due to the impacts of objects coming from the outer parts of the solar system, which instead are deflected by Jupiter.

On the problem of the absence of evidence of the existence of extra-terrestrial civilisations, there is the famous *Fermi paradox*. On a beautiful summer day in 1950, Enrico Fermi and Edward Teller were going to lunch and chatting about space travel-related topics. In the discussion, Teller reports that Fermi asked the following:

> *What I thought about it, and how likely I thought it was that within the next ten years we would observe a material object moving faster than light. I answered 10^{-6}, and Fermi said that was too low a probability. According to him it was more than ten percent. A few minutes later, while we were having lunch and talking about something completely different, Fermi came up with the question "**So where is everyone?**", which provoked general laughter because, although the sentence was totally taken out of context, we all understood that he was talking about extraterrestrial life.*

This question went down in history as the *Fermi paradox*: if the Universe contains an enormous number of galaxies, stars, and planets (which were not known at the time the events took place) and probably advanced civilisations, why do we have no proof of their existence?

Various solutions to the problem have been proposed. Stephen Webb has enumerated 75 possible explanations, which can be divided into several classes:

- The number of intelligent civilisations is small, and in particular, f_i is small.
- Extraterrestrial civilisations exist, but we have no proof of them, i.e. f_c is low. One reason may be that these civilisations are very far away or they transmit signals for a short time.
- The life of intelligent civilisations is short-lived (small L). According to Drake, many intelligent civilisations existed, but they tend to disappear rapidly due to adverse natural events or the self-destruction typical of advanced civilisations.

Ultimately, answering the question of whether we are alone or not appears to be much more complex than it might seem at first glance. The size of the galaxy, the limit of the propagation of electromagnetic signals, and the many effects that lead to the appearance and disappearance of an intelligent civilisation make the problem of knowing the existence of intelligent life forms a very complex problem.

Another problem linked to the Fermi paradox is the possibility of communicating with these civilisations. If we want to be optimistic and assume that there are 1,000 advanced civilisations in our galaxy, we can estimate the average distance between two civilisations. To do this, we recall that our galaxy is a spiral galaxy, with a diameter of 100,000 light years and a thickness of around 1,000 light years. We can approximate the volume of the galaxy with that of a cylinder, obtaining a volume of around 7.9×10^{12} light years cubic. Since we supposed that in the galaxy there are 1,000 civilisations, this means that each one is located in a volume of 7.9×10^{9} light years cubic. The cube containing a civilisation has a radius equal to the cube root of its volume, namely 1,992 light years. This makes us understand why it is very difficult to communicate with these civilisations. By sending a signal, we would receive a response after 3,984 years. Let's not even talk about the possibility of coming into contact with them. If the number of civilisations was equal to 10, the average distance between them would be 9,244 light years, and so on. Summarising, even with a large number of civilisations in our galaxy, we would not have chances to contact them. Other reasons why we have not received signals from extraterrestrial civilisations have been proposed. One possibility is that evolved civilisations have a short life. Another hypothesis is that they are not interested in communicating. The "dark forest hypothesis", according to which several extraterrestrial civilisations exist, but they do not communicate because they consider the other civilisations a threat. Another possibility is that extraterrestrial civilisations use different channels for communications (neutrinos, gravitational waves, or quantum correlations). The previous discussion regards our galaxy. We could extend it to the visible Universe and ask ourselves if in our Universe exist intelligent life or less. We should repeat the chain of reasoning followed until now, recalling that in the observable Universe, there are hundreds of billions of galaxies. We know that at least in one galaxy (our galaxy), there is intelligent life, and because of the large number of galaxies, the probability that there is at least another galaxy with intelligent life is probably not zero. Amir Aczel, in *Probability 1: Why There Must Be Intelligent Life in the Universe*, arrives at the conclusion that the probability of other life in our Universe is 100% using arguments drawn from maths, cosmology, and biology. Cosmology tells us that our Universe is flat, and it is possible that it is infinite. If this were true, we are certain we are not alone in the Universe.

Appendix 1: Big Slurp

We have seen that the *quantum vacuum*, unlike what is commonly understood as a vacuum, is the state of minimum energy of a physical system. However, a system can be trapped in a state with energy higher than the minimum one; that is, it can find itself in a *false vacuum*. The system is therefore not stable. If the potential of the Higgs field is as shown on the left-hand side of Figure A1.1, the field will evolve to a minimum and stop. If the situation is as shown on the right-hand side of Figure A1.1, the yellow line indicates that there is a second lower energy minimum. So, the field could evolve in some way (for example, through *quantum tunnelling*[1]) and "slide" to reach this minimum. The transition from a higher-energy vacuum to a lower-energy vacuum is called *vacuum decay*. If this happened at any point in the universe, the *bubble* of the *new (true) vacuum* would expand at the speed of light, changing the characteristics of our universe or destroying everything. This is unlikely to happen because it requires a huge amount of energy. Furthermore, the time required for this to happen is very long. According to a 2018 study by Andreassen and collaborators, the time needed would be greater than 10^{58} years.

---

[1] The tunnelling effect is a quantum mechanical effect that allows a microscopic object in the quantum world to be in a state that classical mechanics forbids. For example, in the classical world, a projectile that does not have enough energy will not be able to pass through a wall, while this can happen to a particle in the quantum world.

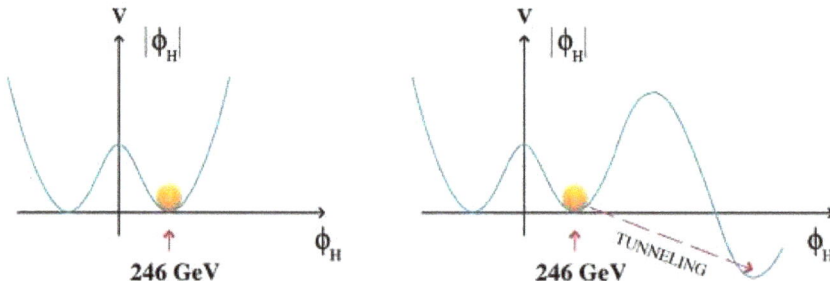

Figure A1.1.   Left: Higgs field at the minimum of its potential. Right: If there is another minimum in the Higgs field potential, the Higgs field can transition to the lower-energy state.

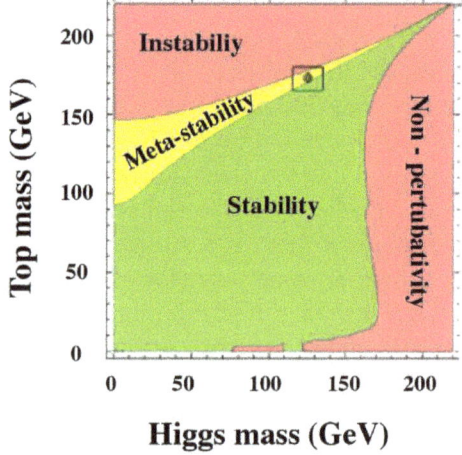

Figure A1.2.   Diagram of the stability of the Universe based on the mass of the top quark (vertical axis) and that of the Higgs boson (horizontal axis). The values of the masses of the two particles tell us that the Universe is in a metastable zone, i.e. one of precarious equilibrium.

Source: Adapted from Degrassi *et al.*, *J. High Energ. Phys.* 98 (2012). https://doi.org/10.1007/JHEP08(2012)098.

This way of ending the new universe is referred to as the *big slurp*.

After determining the mass of the Higgs, it was clear that our universe is in a *metastable state* (Figure A1.2), that is, halfway between stable and unstable.

This means that the big slurp scenario is one of the possible ends of our universe.

What happens after the big slurp is entirely speculative. One possibility, extending some of Caroll's 2004 results, is that if the field that generated inflation still existed in a vacuum, it could reproduce itself.

Appendix 2: Inflation

Guth's Inflation

There are various inflation models, but almost all of them agree on the hypothesis that inflation is driven by an unknown scalar field (similar to the Higgs field) called the *inflaton*. When we described inflation in Chapter 10 (The Multiverse of Chaotic Inflation), we said that the Universe was initially in a state of *false vacuum*. To explain what is meant by a false vacuum, let's consider Figure A2.1.

We indicate the energy state of the Universe with a ball. The Universe can be in state 1. This is an unstable energy minimum. In the same figure, we observe that there is another state, state 3, a minimum that has lower energy and is stable. State 1 is referred to as a false vacuum because it is not the lowest energy minimum. Instead, state 3 is the lowest theoretically conceivable level and is the "true" vacuum. So, for Guth, our Universe was initially in a state similar to state 1, a false vacuum. In quantum mechanics, it is possible for the Universe to transit from the false vacuum of state 1 to the true vacuum of state 3 through a process called the tunnelling effect, as shown in Figure A2.2. Before moving forward, it is better to clarify what quantum tunnelling, or the tunnelling effect, is. In classical mechanics, due to the principle of conservation of energy, a particle cannot overcome an obstacle if it does not have enough energy. For example, if we take a tennis ball and hit it against a wall, i.e. a barrier of potential energy, the ball will reverse its motion and go back. In quantum mechanics, a particle, which has both wave- and particle-like behaviour, has a non-zero probability of crossing an arbitrarily high potential

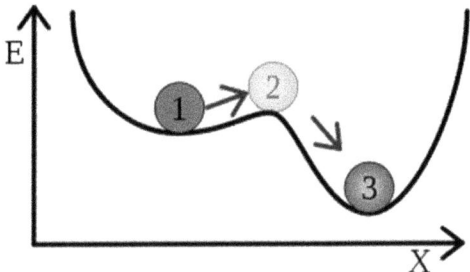

Figure A2.1.   Collapse of the false vacuum. State 1 is a local minimum of energy and constitutes a false vacuum. State 3 represents the lowest possible energy state and is a true vacuum. A transition from the false vacuum to the true vacuum is possible through a quantum mechanical effect called the tunnelling effect. The transition from state 1 to state 3 constitutes the collapse of the false vacuum.

Source: Wikipedia.

energy barrier. The particle is associated with a wave function, the squared modulus of which represents a probability: in our case, the probability of finding the particle in a given region of space. If we consider a one-dimensional potential barrier, the solution to the Schrödinger equation[1] inside the barrier is a decreasing function that never becomes zero. There is, therefore, a certain probability that the particle crosses the barrier. As shown in Figure A2.2, what follows is a phase of fall of the field from the false vacuum to the true vacuum and a final oscillation of the field into the field minimum.

Guth's inflation theory has problems determining the end of inflation. The decay of the inflationary field, the inflaton, follows the rules of quantum mechanics. The onset is not predictable, and the decay occurs at different times in different places in the form of bubbles of true vacuum within the false vacuum. The energy released by the decay is concentrated on the surfaces of the bubbles, and the collision between bubbles would release energy in the form of heat, in a chaotic process that would produce the same amount of inhomogeneity that inflation is supposed to eliminate.

---

[1] The Schrödinger equation is a quantum mechanical equation that allows us to determine the temporal evolution of the state of a system, such as a particle or an atom. The unknown of this equation, called the wave function, is linked to the probability of finding a particle in a given spatial region.

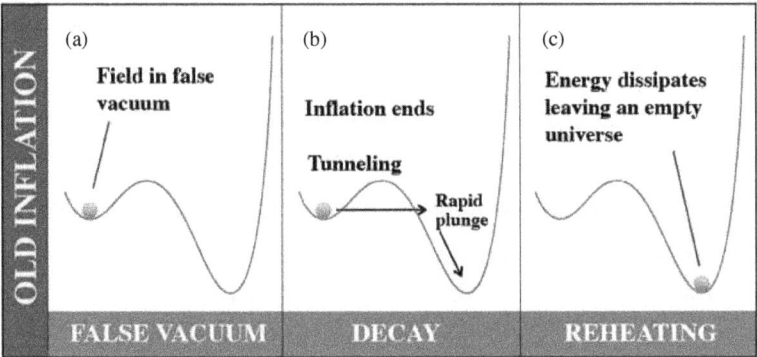

Figure A2.2.　Description of Guth's inflation.

The problem was studied by several cosmologists who proposed a scenario in which the bubbles did not collide but expanded to dimensions larger than those of the observed Universe. This led to variants of Guth's inflation, such as *eternal inflation.*

Other Models for Inflation

Guth's model was replaced by another (by Linde and Steinhardt) in which inflation was slower and exponential expansion does not occur when the field is trapped in the false vacuum. However, the idea shared some commonalities with Guth's inflation. It was assumed that there existed a field similar to that of the Higgs or that of the quintessence, a scalar field called the *inflaton,* and that the Universe was initially in a state of false vacuum in this field. This field was the only one present at the time, distributed throughout space similarly to the Higgs field and with an energy profile (potential energy) shaped similar to the Mexican hat shape of the Higgs field. The initial part of the field profile, corresponding to the false void, is flat, as can be clearly seen in the figure. The trapping of the field in a sort of false vacuum minimum is not observed. So, the tunnelling typical of Guth's model was no longer necessary.

From the false vacuum state, i.e. from the maximum of the curve, the field would have undergone a slow rolling movement (Figure A2.3) towards the true vacuum, the minimum of the curve, i.e. the region of minimum energy, characterised by repulsive gravity. This would cause a

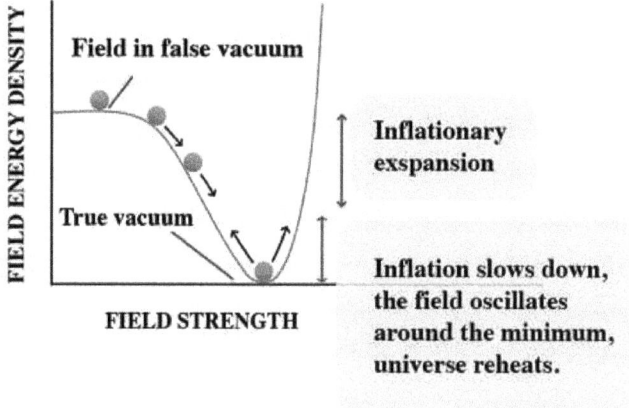

Figure A2.3.   Evolution of the inflationary field.

rapid expansion of space. The field would then continue its rolling phase towards the true vacuum (the minimum of the curve). The energy at the top, in the false vacuum, is greater than that in the minimum area, the true vacuum. When the field reaches the minimum of its potential, in the true vacuum state, it begins to oscillate around it, like a ball dropped down a slope with the shape of the potential profile as shown in the figure. In this way, it would begin to dissipate the energy it had at the top of the curve. The release of energy, according to the dictates of quantum mechanics, produces fields and particles linked to them in the so-called *reheating phase*. Today's particles therefore originated in this phase of the evolution of the *inflaton* (Figure A2.3). Having reached the minimum of potential, i.e. the true vacuum, the pace of expansion would have slowed down to today's rates.

In summary, the Universe was filled with the inflaton field, which was in a state of false vacuum. Quantum fluctuations brought the Universe out of the false vacuum state, pushing it towards the true vacuum state. This initially happened slowly, in the slow rolling phase, in which the Universe was enormously expanded from dimensions of the order of 10^{-28} m up to a few tens of centimetres in the time it took the field to reach its minimum potential. This dimension is the observable Universe.

According to Guth, the size of the Universe could be 10^{23} times larger than that of the observable Universe, but there are very different estimates, both larger and smaller. At this point, fields and particles were generated, and the region in which the inflation occurred began to expand at today's rate.

Like all fields, the inflaton field is subject to quantum fluctuations. These fluctuations cause the potential of inflation to be subject to uncertainty, and it follows that inflation cannot end everywhere at the same instant because the fluctuations impart different values to the field even in two very close regions. So, inflation would end at different times at different points. At points where quantum fluctuations exceed a certain threshold, inflation will occur, and at points below the threshold, it will stop. A bubble of energy and matter would be created, which would give rise to a new universe. Once inflation begins, there will always be a region of exponential expansion that will give rise to another universe. This is one of the variants of inflation, *eternal inflation*, according to which, in different regions of the Universe, the expansion continues forever.

Index